THIS IS
YOUR CAPTAIN
THE NAKED TRUTH ABOUT
THE PERSON FLYING YOUR PLANE

By Captain Jack Watson

THIS IS YOUR CAPTAIN: THE NAKED TRUTH ABOUT THE PERSON FLYING YOUR PLANE

Library of Congress Cataloging-in-Publication Data

Watson, Jack
  This Is Your Captain: The Naked Truth About The Person Flying Your Plane / by Captain Jack Watson.
  p. cm.
  Includes bibliographical references and index.
  ISBN-13: 978-1-62023-102-9 (alk. paper)
  ISBN-10: 1-62023-102-6 (alk. paper)
1. Air pilots. 2. Air pilots--Psychology.
3. Airplanes--Piloting. 4. Airplanes--Piloting--Human factors. 5. Leadership. 6. Watson, Jack, 1947- I. Title.
  TL561.W38 2015
  629.13--dc23
                              2015011938

Printed in the United States

Printed on Recycled Paper

Reduce. Reuse.
RECYCLE.

A decade ago, Atlantic Publishing signed the Green Press Initiative. These guidelines promote environmentally friendly practices, such as using recycled stock and vegetable-based inks, avoiding waste, choosing energy-efficient resources, and promoting a no-pulping policy. We now use 100-percent recycled stock on all our books. The results: in one year, switching to post-consumer recycled stock saved 24 mature trees, 5,000 gallons of water, the equivalent of the total energy used for one home in a year, and the equivalent of the greenhouse gases from one car driven for a year.

Over the years, we have adopted a number of dogs from rescues and shelters. First there was Bear and after he passed, Ginger and Scout. Now, we have Kira, another rescue. They have brought immense joy and love not just into our lives, but into the lives of all who met them.

We want you to know a portion of the profits of this book will be donated in Bear, Ginger and Scout's memory to local animal shelters, parks, conservation organizations, and other individuals and nonprofit organizations in need of assistance.

– Douglas & Sherri Brown,
President & Vice-President of Atlantic Publishing

DEDICATION

To Airline Pilots everywhere —
Thank you for giving
so much of your lives to
bring the world closer together.
For all the Flight Attendants —
You are appreciated more
than you'll ever know.
Thank you for watching over all of us.

TABLE OF CONTENTS

ACKNOWLEDGEMENTS

This is my seventh book for Atlantic Publishing. Publisher Doug Brown and his lovely wife Sherri have as always been so gracious with their time and encouragement, having them in my corner is appreciated more than words can say. From the bottom of my heart, thank you both.

To my favorite editor Melissa Figueroa goes a special thanks for allowing an old captain to tell his story. Your candor and humorous comments are most appreciated. I hope everyone enjoys this story as much as you.

Once again the incredibly talented Jacqueline Miller pieced my words together with captain-like precision, not to mention creating the dramatic cover art. As always Jackie, many thanks (you're the best dancer and book designer I know.)

A special thanks to equally talented graphics designer Meg Buchner who provided finishing touches and technical advise throughout the printing process.

All of my books are available as an electronic download in addition to printed versions. This feat of magic was brought to you by the talented Linda Hambright, Atlantic's e-book specialist. Thanks Linda for your continued assistance.

One of the key people at Atlantic is the very organized and always helpful Crystal Edwards. Without her solutions and knowledge I doubt if I could have ever finished my first book. Thank you Crystal — Joey says "Meow."

To my many retired and active captain friends, over the years you've unwittingly provided me with the gift of your wisdom. These stories helped form my thoughts for this book. I am honored to have known you all.

FOREWORD

This is a book for dreamers, voyeurs, passengers and pilots. Listed as one of the top three most deadly professions in America, the airline pilot deserves equal amounts of respect and admiration.

Who are these men and women that strap on several hundred million dollars of aluminum, electronic and other exotic materials, then launch skyward to cruise 9 miles above Earth traveling 500 miles per hour? More specifically, who is the captain— the person you'll probably never meet but by random chance have entrusted your life to?

I'm certain some pilots within the airline industry will disagree with my conclusions and observations, I accept that fact; this is inevitable.

Based on my almost 40 years of observation, there seems to be an infinite number of ways one can end up in the captain's seat of an airliner. What's portrayed here is, for lack of a better term, a generic climb through a seniority list to the captain seat. This example journey is based on countless pilot career stories shared with me over the years.

Let me be clear — this is not a story about death defying adventures of wacky daredevils. Men and women who captain large jets for major air carriers are by and large among the most stable individuals on Earth. By the time they reach the lofty position of captain they've most likely been vetted more than a U.S. Supreme Court Justice. The following chapters detail a typical airline pilot life. As you'll see this is anything but a normal occupation. The upside for those choosing this profession is that one day they'll sew four stripes on their coat sleeves — quite literally be given the keys to a jet costing hundreds of millions of dollars, and be sent off without direct supervision into a crowded sky, making millions of dollars for an airline as a high-flying bus jockey.

But, unlike bus drivers that can pull over at the first sign of trouble, airline captains do not have that luxury. The penalty for handling even a minor problem incorrectly can easily lead to their death plus hundreds — even thousands of others. Still interested? Then read on — You're about to enter the world of "YOUR CAPTAIN."

Captain Jack Watson - Retired
March 2015

Author's Note

There are many second and third tier carriers flying turboprop and midsize jets — captains at this level are not the focus of this book.

In the airline industry pilots flying for large legacy carriers like United, Delta, American, etc., are looked upon as the top-of-the-pyramid pilots. Flying for these gigantic carriers usually means bigger equipment, more diverse routes, better selection of domiciles, and most importantly to its pilots — larger paychecks. Captains are a unique breed at legacy carriers — this is their story.

CHAPTER 1
In The Beginning

My dad used to say, "You can't do the job unless you understand how the tools work." Knowing his wisdom I suspect he would have also said; in order to fully understand the complexities of being an airline captain it's necessary to understand the evolution of the industry that created this position.

While other countries may operate airlines in a similar fashion, in my humble opinion the gold standard for operational expertise is United States certified air carriers. What follows is a brief history of airline development in America.

* * *

Under auspices of the U.S. Post Office, airmail operations began in 1918 to hopefully stimulate airplane production and generate a pool of reliable and qualified pilots.

Using World War I trainers converted to mail planes, early airmail service floundered, especially when weather turned bad.

After the war, larger airplanes were used allowing mail planes to beat train delivery times over longer distances.

By 1924, coast-to-coast airmail service used light beacons to guide open-cockpit planes at night. Mail from New York now arrived on the west coast in two days instead of the usual five by rail. This time savings had a direct impact on expediting check clearing, and other similar time sensitive business paper.

As airmail delivery became a workable system it established a considerable following, forcing the Post Office into yielding to congressional pressures. Trains were vastly slower and now an unwanted means of mail carrying. As a consequence the Contract Air Mail Act of 1925 turned over the mail service to private contractors. A year later, the Air Commerce Act established a bureau to enforce procedures for licensing of aircraft, engines, pilots, and other personnel. This act stimulated design and production of more advanced planes. Additionally the act reassured insurance companies, investors, and banks that safety standards would be enforced.

The Air Commerce Act was a turning point for America and prompted rapid growth in the burgeoning aviation industry. The United States had taken the lead globally by turning over civil aviation to commercial operators, allowing aggressive competition that accelerated major developments in aviation technology and aircraft performance. What followed was development of lighter and more powerful engines and streamlining of aircraft designs that increased overall performance.

In the late 1920s and early 1930s, the U.S. Post Office introduced pay formulas for airmail carriers that favored larger aircraft that could accommodate passengers as well as mail. Aeronautical engineers produced volumes of research that improved aircraft designs. The result was an increase in development of larger planes that lead to development of tri-motor monoplane transports marketed by a subsidiary of Ford and by the European aircraft manufacturer Anthony Fokker, who had set up shop in the United States.

The Ford Tri-motor was one of the first cabin class airline transports in America. Sporting three radial engines it was reliable, controllable with one engine out, and, more importantly, spurred development of larger, safer and speedier airliners in the future.

American technology took a major step forward with the introduction of the Boeing Company Model 247 airliner, which cruised at about 180 mph and entered service with United Airlines, Inc., in 1933. With its all-metal stressed-skin construction, retractable gear, two 550-horsepower Pratt & Whitney Wasp radial engines, the 10-passenger airliner seemed to be head-and-shoulders above competitive aircraft.

A short time before the 247 began flying, a Fokker tri-motor of Transcontinental & Western Air,
Inc. (TWA) crashed. Everybody aboard died, including famed University of Notre Dame's football coach
Knute Rockne. Investigation of the crash raised
questions about structural weakness in the plane's
wooden wing spar. That controversy gave wood-
en-wing spar tri-motor airliners a bad reputation.
When TWA asked manufacturers to submit designs
for a replacement, Douglas Aircraft Company sub-
mitted their DC-2 design of an all-metal twin-engine airliner. With many advanced improvements
and a 14-passenger capacity, it surpassed Boeing's
Model 247 submission.

Here's an interesting side note about the Boeing 247 — it was the victim of the first proven explosive device to destroy an airliner. A United Airlines Boeing 247 exploded over Chesterton, Indiana on October 10, 1933. Regrettably this accident probably ushered in the concept of using the aircraft itself for nefarious purposes.

Following success of the DC-2 the legendary DC-3 entered service in 1936. It cruised at 185 mph, and carried 21 passengers—double the capacity of the Boeing 247. By 1939, with superior seat capacity, and performance, DC-3 transports were carrying 90 percent of the world's airline traffic.

Douglas transports dramatically improved air travel within the United States, but airline entrepreneurs were looking for airliners capable of transoceanic travel. Many in the 1930s still believed huge gas-filled airships would be the key. American Airlines, Inc., publicized special schedules that allowed DC-3 passengers to make transatlantic connections with the *Hindenburg's* airship terminal in Lakehurst, New Jersey.

In 1937 as the highly volatile Hydrogen gas filled Hindenburg was on approach for landing at Lakehurst it suddenly exploded, ending this short-lived arrangement. Plans for utilizing dirigibles as passenger liners were soon forgotten.

The original flying boats of Pan American World Airways, Inc. (Pan Am), were designs from Russian-born American engineer Igor Sikorsky. Pan Am operated them over water in the Caribbean region, saving weeks of travel time compared with steamship and awkward railway connections.

By late 1930, American manufacturers like Martin Company (now the Martin Marietta Corporation), Boeing, and Sikorsky were all producing large four-engine flying boats intended for flights over the Atlantic and Pacific.

By 1935, using islands strung across the Pacific, Pan Am completed construction of many mid-ocean stopover facilities and had installed its own radio communications and meteorological network. Using Martin flying boats, most of the early Pan Am flights across the Pacific carried mail, along with an occasional government or business passengers who could afford the high fares.

In 1939 the imposing Boeing 314 Pan Am Yankee Clipper flying boat began scheduled airmail and luxury passenger service across the Atlantic to Europe. These promising and expensive flights were soon curtailed by wartime conditions in Asia and Europe.

In the early 1940s, large sparsely populated areas separated the United States. But as America grew, larger cities blossomed across the landscape with intervening distances that made airline service a desirable means of connecting this vast and growing city network.

Airline transport designs in America favored speed over comfort for its passengers. As anyone who has flown recently can attest — this desire for speed over comfort still exist, but that's probably dictated by dollar conscious airline executives watching bottom lines for the sake of their shareholders.

Boeing's pressurized cabin Stratoliner entered service in 1940. Pressurization allowed airliners to fly above a significant amount of adverse weather. This greatly improved on-time performance schedules and usually gave passengers a more comfortable ride. Moreover, flying at higher altitudes increased fuel efficiency, thereby increasing range. Only a small number of Stratoliners entered service before World War II, which led Boeing to focus on building bombers.

To counter Boeing's Stratoliner, Douglas introduced the DC-4. Although unpressurized, it had comparable performance to the Stratoliner and, most notably, could carry more passengers. The DC-4 had a tricycle landing gear, unlike the Stratoliner's conventional tail wheel. The tricycle gear improved the pilots' view especially during taxi and takeoff, and enhanced the plane's takeoff performance.

The DC-4 produced as the C-54 during the war was the U.S. Army Air Corp's principal long-range transport.

Late in the war, the C-54 was joined by Lockheed's L-1049 Constellation — instantly identifiable by its triple vertical fins.

Originally designed in 1939 as a commercial airliner the "Connie," as it was affectionately known, blended a pressurized fuselage, tricycle landing gear and other advanced technologies.

Development of airliners and scheduled air travel relied heavily on the evolution of aeronautical science and research. The NACA (National Advisory Committee for Aeronautics) established in 1915, evolved into one of the world's leading aeronautical research centers. Additionally, the creation of specialized organizations to investigate accidents, determine probable cause, and make recommendations to avoid accident repetition played a key role in improving air travel safety.

On September 24, 1929, Lt. James Doolittle made the first instrument takeoff, flight and landing at Mitchel Field in Garden City, New York. His responses to a combination of electronic signals and airplane instruments permitted the first successful "blind flight." This experiment represented a monumental lunge forward for all of aviation. It meant piloting an airplane without reference to the ground would be possible using only the instruments in the cockpit for attitude reference and navigation. As for Lt. Doolittle — he would later lead a famous raid over Tokyo, Japan during WWII, receive the Medal of Honor, and retire a four-star general in the USAF.

As 1930 approached, most major cities in the United States had established municipal airports. During the great depression following "Black Tuesday," on October 29, 1929, various New Deal construction programs improved and built additional airfields across America. All-weather paved runways would soon replaced sod and dirt landing areas. Federal funding and assistance allowed major airfields to acquire control towers and radio equipment integral to a workable air traffic control system. The stage was set for the modern airliner.

During WWII the United States supplied most of the air transports for Allied forces. Drafted into military service, the C-47 (DC-3) and C-54 (DC-4) became workhorses for the U.S., and its allies.

After the war development of postwar airlines was inevitable. In 1944, a historic meeting convened in Chicago where international representatives agreed on a provisional administrative entity to govern civil aviation. By 1947, the International Civil Aviation Organization (ICAO) settled in Montreal as an adjunct of the new United Nations organization. The new organization specified English as the universal language for pilots and air traffic controllers engaged in international operations. Additionally they specified standardized formats for radio frequencies, terminology, emergency procedures, navigational equipment, runway markings, airport lighting, etc. These mandates would prevent a chaotic, and unacceptably dangerous evolution as air travel became global in reach.

After the war, many United States airlines looked for DC-3 replacements to use on short-haul and medium range flights. Consolidated Vultee Aircraft Corporation, (Convair) built the twin-engine 240/340/440 series, and sold more than 1,000 models between 1947 and 1956. Several hundred military versions also filtered back into civil use. Convairs could cruise at 280 miles per hour and their pressurized cabin could provide environmental comfort for 40 to 50 passengers depending on the model. Subsequent turboprop conversions like the Convair 580 and 600 kept the type in service for several decades.

After 1945, Douglas introduced the pressurized DC-6 to challenge Lockheed's Constellation on its dominance of domestic and international routes. American manufacturers were constantly challenging one another to improve their product line. North America's market for airliners generated high-volume production, keeping unit cost low. Eventually, the value of postwar American designs led to a dominating presence in airline fleets globally.

Boeing was envious of Lockheed, Convair and Douglas' position in the airliner market place — they wanted in. The Stratocruiser was their answer. It offered unmatched luxury for air travelers in the late 1940s and early '50s. Its famously spacious cabin seated 55 passengers, and its bar/lounge, entered through a spiral staircase to the lower deck, created a sensation. Pan Am quickly introduced Stratocruisers on premier routes across the North Atlantic. Eventually, even the Stratocruiser fell from favor against the faster piston-engine airliners from Douglas and Lockheed.

Transcontinental flights in the United States invariably required a stop for fuel en route. These constraints would slowly disappear in the 1950s with the Lockheed Super Constellation and the Douglas DC-7. The final version of the DC-7 appeared between 1956–57 and was known as the DC-7C, or "Seven

Seas," by those who flew her. She was capable of nonstop transatlantic flights in either direction — the equally capable Lockheed 1649A Starliner could fly nonstop on polar routes from Los Angeles to Europe. The Starliner carried 75 passengers cruising at speeds up to 400 miles per hour. Each of its radial piston engines developed 3,400 horsepower. Prior to jet transports, both of these workhorse aircraft transformed the realm of air travel and remained in service with major airlines late into the 1960s.

As early as 1953 United States domestic airlines reported more passenger miles than railroad Pullman travel. By that years end airlines had taken the lead as prime mover for American travelers making trips greater than 200 miles. By 1958 the majority of U.S. passengers headed for Europe chose to fly rather than travel by ocean liner.

Jet engines evolved at a time when reciprocating engines and propellers were reaching their functional limits. A jet engine was mechanically superior to a reciprocating engine. These engines developed rapidly and by 1950 they were producing thrust that was impossible to match with piston engines. Reciprocating aircraft engines reached a practical limit with the 3,500-horsepower, 28-cylinder Pratt & Whitney R-4360 engine. Today a modern jet engine like General Electric's GE90-115 used on the Boeing 777 can produce as much as 115,000 pounds of thrust (over 100,000 horsepower.)

Two stages of jetliner development marked the 1960s. First was adoption of the turbofan engine. The second stage was marked by the 1969 introduction of the wide-body, 400-seat Boeing 747.

Boeing had demonstrated its mastery of producing aircraft to meet new demands by is creation of the three-engine T-tailed 727 in 1963, followed by a succession of aircraft (the 737 [1967], 757 [1982], 767 [1981], 777 [2003], and 787 [2007]) each aircraft design was tailored to the needs of specific airlines and routes. Boeing's 737, with more than 6,000 sales, is the most produced jet transport in history.

During Boeing's expansion, Douglas had management problems. While its DC-9 was a spectacular success, they were unable to match Boeing's proliferation of designs. McDonnell Aircraft Corporation acquired Douglas in 1967. The new McDonnell Douglas Corporation produced the tri-motored DC-10 to meet an estimated market requirement for about 750 wide-body fuel-efficient aircraft.

Lockheed tried to enter the same market with its technologically advanced L-1011 TriStar. McDonnell Douglas sold 446 DC-10s. Lockheed sold 250 TriStars. Both companies lost massive amounts of money. McDonnell Douglas struggled to prevail with the MD-11, an improved DC-10. They continued producing the DC-9 as the MD-80 and MD-90 series. When Boeing acquired the firm in 1997, it applied the designation 717 to one version of the former DC-9 twin-engine jet.

In spite of the intense competition to build jet airliners, a new entrant appeared in the early 1970s named Airbus Industries. It was co-owned by a consortium of French, German, British, Spanish, Dutch, and Belgium companies. It subcontracted many parts to other countries. Discounted at first as having little chance to compete, its aircraft lineup eventually became widely accepted. Complaints by United States aircraft manufacturers that Airbus was being subsidized by the various European governments was countered by Airbus charging American manufacturers for subsidized sale of their aircraft to U S military and other government agencies. Accusations from both sides were true.

Airbus was the first manufacturer to introduce fly-by-wire technology to an airliner cockpit with introduction of the A-320. This airplane marked the latest evolution of airliners into commercial service. Flight control laws (algorithms) on the A-320 converted pilot control inputs into code that actuated computer operated flight control surfaces and other aircraft systems. No longer was there a need for the majority of levers, cables, push rods, and pulleys to connect the pilot's controls to the airplane's control surfaces or systems. This type of aircraft control was a first for commercial aviation (other than military. It is now the gold standard of aircraft control.

The first digital fly-by-wire concept was in use as early as 1969 on the Lunar Module that landed on the moon. Later the F-16 and F-117 Stealth fighter employed fly-by-wire technology.

I had the good fortune to be an instructor pilot and captain on the A-320 for several years and have high praises for the technology. It was a giant leap technically for my 'old school' brain to adjust to the concept of controlling a computer driven jet. It is a complex airplane, without doubt. Understanding how it worked required learning a new language.

The following paragraphs are extracts from an Aérospatiale report on Electrical Flight Controls. It is one of many papers released by Airbus and others that discuss fly-by-wire systems. It's educational to understand why this new technology is growing in popularity.

Weight savings, redundancy and a new way of thinking are just a few of the benefits and challenges pilots face in this new frontier of mostly automated flying.

If you think the report is confusing, imagine what the first classes of old airline captains, used to mechanical airplanes, were thinking when they sat in the initial ground schools on the A-320. This was "Buck Rogers" stuff, and most of us ate it up, but a few washed out.

"The overall dependability of the aircraft fly-by-wire system relies in particular on the computer arrangement   (the   so-called   control/monitor architecture), the system tolerance to both hardware and software failures, the servo-control and power supply arrangement, the failure monitoring, and the system protection against external aggressions. It does this without forgetting the flight control laws, which minimize the crew workload, the flight envelope protections which allow fast reactions while keeping the aircraft in the safe part of the flight envelope, and finally the system design and validation methods.

The aircraft safety is demonstrated by using both qualitative and quantitative assessments; this approach is consistent with the airworthiness regulation. Qualitative assessment is used to deal with design faults, inter- action (maintenance, crew) faults, and external environmental hazard. For physical ("hardware") faults, both qualitative and quantitative assessments are used. The quantitative assessment covers the FAR/JAR 25.1309 requirement, and links the failure condition classification (minor to catastrophic) to its probability target."

When I first got on the Airbus, like most of the pilots in my class, the above text would have read like a foreign language. Strangely, after flying and instructing on the A-320 for several years I actually understand what these paragraphs mean. I won't lie to you, the learning curve was steep. Checking out on the A-320s right after they were introduced was like preparing for the first manned mission to space. Well, maybe not that difficult. But there was a huge washout rate at the beginning of the program. It wasn't that the new 'fly-by-wire' pilots were slow learners, it was simply a problem with determining what was important to teach.

There are approximately 230 computers in an A-320. When the FAA was initially tasked with giving ratings in the A-320 they reasoned (incorrectly) that we should know about every nuance of the airplane. As it turns out, that was for the most part a waste of a pilots' brainpower.

The beauty of a computer driven airplane is that its computers make a lot of decisions without pilot input. What we see on the cockpit 'glass' displays is what the computers want us to see. The automation is designed to reduce pilot workload. Bottom line, we found early on it was only necessary to teach what we could actually control ourselves.

The A-320 training philosophy at my airline was adjusted to reflect 'teach what they actually needed to know' concepts. Just because you can fly an airplane doesn't mean you have to know how to build one. Because of this initially new technology, the first A-320 pilots were tested on just about everything, since no one knew for sure what knowledge was, and was not necessary to pilot one of these modern marvels. Those 'know everything' days are thankfully behind us.

The human brain is amazing, but it's no match for the unbiased reasoning of a well-programmed computer.

From the first airmail flights to computer-operated airliners nearly a century has past. What lies beyond the horizon is anyone's guess. Our limits on speed and capacity are both technical and psychological. How big is too big, and how fast is too fast? We are at a new genesis in air travel. The world's largest airliner as of 2014 is the Airbus A-380. It cost over 400 million a plane and can carry up to 853 people in an all economy class configuration. It is an amazing aircraft, with 50 percent less engine noise than a Boeing 747-400 on takeoff. Is this the future?

Our global airspace is becoming overcrowded with airliners. Bigger jets might mean a reduction in that congestion. It could also mean fewer pilot jobs.

For those contemplating a career in the cockpit, fear not: at almost a half billion dollars each it will be many, many years before the crowded skies saturated with behemoths like the A-380.

CHAPTER 2
Job Description

From the early years of airmail until present day the primary roll of an airline pilot-in-command (PIC or captain) has changed very little. Protect those who entrusted their lives and property to you and don't bend the airplane — sounds relatively easy. Seventy-five years ago it was, especially when all you might be flying is a bag of mail and a few small parcels. Like any industry, as air travel gained popularity a captain's job description became more complex.

A modern day captain is usually the end product of years of training, watching and waiting — with emphasis on the waiting.

Major airlines in the United States employ thousands of pilots with a companies pilot seniority list often numbering in excess of 10,000 aviators. Describing the captain's job without mentioning the pilot seniority list would be telling an incomplete story. This is the most important list in an airline pilot's career.

Why are some pilots captains and others first officers or flight engineers? It's all based on seniority. For the most part, ability or expertise in flying has nothing to do with it. Let me elaborate.

Pilots for all major carriers typically work under a contract negotiated by a union, or in-house pilot organization acting as a voice for its respective pilot group. This in-depth contract generally lays out work-rules and a compensation agreement acceptable to all parties.

The airline's seniority list determines a pilot's climb through the ranks and literally shapes the quality of his or her life. The list is generally a reference part of the contract only, but if the list is an amalgamation caused by a merger between two or more carriers — how the combined list is integrated would necessarily be a larger part of, or addendum to an existing contract.

Integration or merging of seniority lists is usually a reluctant and heated battle between the different pilot groups involved. Management doesn't really care how the list is merged unless it costs them money (additional training cost, etc.) Pilots do care however — they care a lot. Here's why.

Any pilot hiring on with a major airline in the US goes on the bottom of the seniority list. If they are part of a class of new hire pilots that start training on the same day, they are numerically added to the list by age. The youngest person is given the lowest (bottom) seniority number and the oldest, the most (highest) senior number in the class. It is an arbitrary system established decades ago and awards advancement based solely on seniority, and is not based on merit or flying skills. Translated — all pilots are not created equal, and neither is the seniority list.

As can be expected (with any job) some pilots are great at what they do, while others fall short. Worry not. As a paying passenger you can expect a safe level of competence from those seated in the airliner cockpit — some crews are just better than others. Continuous testing and requalification of all aircrews assures each pilot meets a predetermined and acceptable set of standards. To clarify, in an imperfect world some crews are better than others, but again, all of them are capable. The biggest difference (my opinion) is in the experience level. In a way, that's why a seniority system actually works to the advantage of the paying public. Junior equipment, (usually smaller and less complex aircraft) are normally flown by competent crews with lower seniority numbers. They typically have less experience than those flying larger and more complex aircraft. However, the degrees of complexity are narrowing between big and small aircraft as technology evolves. Generally speaking junior pilots spend considerable time flying smaller airplanes on shorter domestic routes as opposed to 'long haul' or international flying that can be quite a bit more demanding. The old "Learn to walk before you run" philosophy is attached to junior flight deck crews.

The captain with a seniority number of one has first choice of equipment and routes they fly and is usually the highest paid and one of the most experienced aviators in the pilot ranks — you'll likely find them in the captain's seat of the carrier's largest jet. The individual with the bottom seniority number conversely gets the least pay, worst trips and flies the most junior (usually smallest) equipment. Adding further to their misery, when first hired, new pilots usually have to sit by a telephone on reserve awaiting a call-out from crew scheduling to report for work sometimes with only an hour notice.

Reserve often eats up 20-22 days a month for a new hire. Depending on the need for reserves, a reserve pilot might not fly at all during the month, but they're still tied to the telephone. Looking back all I can remember about reserve is that I had 'less than' no life. My few days off each month hardly provided enough time to get through the "Honey-do" list.

Like a lot of pilots, I chose to commute to my base, even when on reserve status. That meant a lot of time in a cheap (paid for by me) hotel room near the airport waiting for the phone to ring. Unfortunately for those on reserve, unless hiring continues, reserve status can continue for years. Ask any pilot what they think about being on reserve and the usual response is "It sucks." Enough said.

* * *

A captain's job description has been defined and re-defined by hundreds of Human Resource types, chief pilots, mid level managers, and even our own FAA. I've read many of these inflated scholarly works; some are good, some ridiculous. I've given a great deal of thought to this subject and what follows is an ad hoc opinion about the captain job and what it entails.

Captains are made, not born into the position.
The burden of command falls squarely on his or
her shoulders for all operations of their aircraft.
Unlike an ocean liner captain or a titan of industry,
when critical situations present themselves there is
usually little time for an airline captain to sort out
the problem. Life and death decisions are usually
made in minutes or even seconds. They usually
have little time to discuss a problem with others
aside from their immediate crew. In an ideal world,
airborne problems could be discussed via radio with
people on the ground that can help troubleshoot by
accessing many available and valuable resources.
Unfortunately, due to the immediacy of problems
airplanes can encounter, this seldom occurs during
complex emergencies.

**Resourcefulness and great time management
skills** are traits that don't come naturally to many
of us. Here's where the advantages of a large seniority list can come into play.

All the airline pilots I've known over the years (with
few exceptions) were Type-A personalities. Most, if
not all, were ready, and felt qualified to assume the
position of captain the day they were hired. While
technically and legally most of us could command
an airliner — were we really ready for the job? The
truthful answer is — probably not.

Like a surgeon, whose training is lengthy, pilots only get better with years of experience. It's one thing we unfortunately can't teach.

I spent over a decade in the right seat, enjoying the job but counting seniority numbers until it was my turn to upgrade. I'd leave travel brochures in captain's mailboxes, talk to them about the joy their life might have if they took early retirement, and even encouraged them to take up dangerous hobbies. Seriously — I did all of that! Fortunately for the paying public my lengthy apprenticeship was a bittersweet education that, in retrospect, made me a safer pilot and hopefully a better captain.

Every trip I flew as an engineer or first officer provided me the opportunity to observe both weak and strong captains manage their cockpits. First off, let me apologize for the use of the 'weak' and 'strong' terms. But in reality that's how I'd classify captains I flew with. Over time I found myself adopting mannerisms and management styles of those captains I admired. Conversely I learned how I didn't want to act by watching those in the "weak" column.

Let me be clear. Every captain I flew with was safe — but I wisely kept my fingers crossed that if something went wrong I was flying with a 'strong-column' captain.

Ability to get along is one of the least mentioned but most important aspects of being an airline pilot, especially for captains. The person in the left seat usually sets the mood for how a trip will go. A happy crew is a functional crew. Anything less creates inefficiency and poor communication.

This sounds simple, but the psychology of efficiently commanding an airline crew is far more complex than you can imagine. In the fast paced airline business a crew member bounces around airplanes like a cue ball on a pool table. Because of complex rest requirements, different schedules, etc., it's not uncommon for front and backend (cockpit and cabin) crews to change several times during a schedule of trips.

Each time a crew member is added or taken away the people-dynamic of that particular crew changes. Here's an example. Your start a trip with a first officer who, like the captain, loves hunting, and during the first three or four legs of the trip they both get along famously sharing stories during idle (non sterile cockpit) moments, and working as a cohesive team piloting the aircraft. On the fifth leg of the trip a new F/O replaces the first hunter type. Unfortunately this F/O hates guns and abhors hunting. Hopefully you can see where this is going. A good captain would seek common ground to set the stage for a pleasant trip. A weak captain would turn silent and make the trip unbearable for everyone. This is an oversimplification, but keep in mind when a flight deck crew locks the cockpit door their work environment becomes a very tiny space where personality clashes have to be shelved. You can't just get up and walk away.

Here's a tip. If you want nice guests at a party, invite airline crews — they can usually get along with just about anyone. Plus we like parties!

Technical skills are not only necessary, but also a large part of the actual job of flying. Like most complex jobs, pilots have their own way of communicating on technical subjects. In lieu of large unpronounceable 'doctor-like' words aircraft manufacturers have created large airplane specific languages full of acronyms and abbreviations. Here's an example of terms Airbus pilots know and use to describe their aircraft. It's a long and necessary list, and here's why.

Inflight description of serious airplane problems is usually under a rapidly diminishing time restraint. When airborne, the longer you take to describe problems to people that can possibly help, usually by radio contact with maintenance personnel, the less time you have to find a fix. If you waste too much time the situation might deteriorate to a tragic ending. That's part of the reason why abbreviations and acronyms are necessary. There's no time or reason for use of big words in a cockpit environment. Pilots don't usually have the luxury of unlimited time in dealing with inflight emergencies.

Many of the acronyms listed are actually used by Airbus crews and pronounced as actual words — some are used as shorthand when entering an aircraft logbook mechanical write-up for maintenance to address.

A

ABN - Abnormal

ACARS - ARINC Communications and Reporting System

ACM - Air Cycle Machine

ACP - Audio Control Panel

ACT - Additional Center Tank

ADIRS - Air Data Inertial Reference System

ADIRU - Air Data Inertial Reference Unit

ADM - Air Data Module

ADR - Air Data Reference

ADV - Advisory

AEVC - Avionics Equipment Ventilation Controller

AFS - Auto Flight System

AIDS - Aircraft Integrated Data System

AIU - Audio Interface Unit

AMU - Audio Management Unit

ANP - Actual Navigation Performance

APPU - Asymmetry Position Pick Off Unit

APU - Auxiliary Power Unit

ARPT - Airport

ASAP - As Soon As Possible

ASI - Air Speed Indicator

A/SKID - Anti-Skid

ATE - Automated Test Equipment

A/THR - Auto Thrust

ATS - Auto Thrust System

ATSU - Air Traffic Service Unit

AWY - Airway

B

B - Blue

BARO - Barometric

BCL - Battery Charge Limiter

BCDS - Bite Centralized Data System

BFO - Beat Frequency Oscillator

BIU - Bite Interface Unit

BMC - Bleed Monitoring Computer

BNR - Binary

BRK - Brake

BSCU - Brake Steering Control Unit

BTC - Bus Tie Contactor

C

CBMS - Circuit Breaker Monitoring System

CFDIU - Centralized Fault Data Interface Unit

CFDS - Centralized Fault Display System

CHC - Cargo Heat Controller

CHG - Change

CIDS - Cabin Intercommunication Data System

C/L - Checklist

CO RTE - Company Route

CONF - Configuration (Flaps/Slats)

CPC - Cabin Pressure Controller

CPCU - Cabin Pressure Controller Unit

CRC - Continuous Repetitive Chime

CRG - Cargo

CSCU - Cargo Smoke Control Unit

CSM/G - Constant Speed Motor/Generator

CSTR - Constraint

CTL PNL - Control Panel

CVR - Cockpit Voice Recorder

D

DA - Drift Angle

DAR - Digital AIDS Recorder

DDRMI - Digital Distance and Radio Magnetic Indicator

DFA - Delayed Flap Approach

DIR TO - Direct To

DITS - Digital Information Transfer System

DMC - Display Management Computer

DSDL - Dedicated Serial Data Link

DU - Display Unit

E

ECAM - Electronic Centralized Aircraft Monitoring

ECB - Electronic Control Box (APU)

ECM - Engine Conditioning Monitoring

ECON - Economic

ECP - ECAM Control Panel

ECS - Environmental Control System

ECU - Engine Control Unit

EDP - Engine Driven Pump

EEC - Electronic Engine Computer

EFCS - Electronic Flight Control System

EFIS - Electronic Flight Instrument System

EFOB - Estimated Fuel On Board

EIU - Engine Interface Unit

EIS - Electronic Instruments System

ELAC - Elevator Aileron Computer

EMER GEN - Emergency Generator

EO - Engine Out

EPE - Estimated Position Error

EGPWS - Enhanced Ground Proximity Warning System

ESS - Essential

EST - Estimated

ETE - Estimated Time Enroute

ETP - Equal Time Point

EVMU - Engine Vibration Monitoring Unit

E/WD - Engine/Warning Display

EXT PWR - External Power

EXTN – Extension

F

FAC - Flight Augmentation Computer

FADEC - Full Authority Digital Engine Control

FAP - Forward Attendant Panel

FAV - Fan Air Valve

F/C - Flight Crew

FCDC - Flight Control Data Concentrator

FCU - Flight Control Unit

FD - Flight Director

FDIU - Flight Data Interface Unit

FDU - Fire Detection Unit

FF - Fuel Flow

FGC - Flight Guidance Computer

FIDS - Fault Isolation and Detection System

FLSCU - Fuel Level Sensing Control Unit

FLT CTL - Flight Control

FLX/MCT - Flex/Maximum Continuous Thrust

FMA - Flight Mode Annunciator

FMGC - Flight Management Guidance
        Envelope Computer

FMGS - Flight Management Guidance
        Envelope System

F-PLN - Flight Plan

FPA - Flight Path Angle

FPD - Flight Path Director

FPPU - Feedback Position Pick-off Unit

FPV - Flight Path Vector

FQI/FQU - Fuel Quantity Indication/Unit

FQIC - Fuel Quantity Indication Computer

FRT - Front

FRV - Fuel Return Valve

FT/MN - Feet per Minute

FU - Fuel Used

FWC - Flight Warning Computer

FWS - Flight Warning System

G

G - Green

GCU - Generator Control Unit

GLC - Generator Line Contactor

GNADIRS - Global Navigation Air Data Inertial Reference System

GPCU - Ground Power Control Unit

GRND - Ground

GRP - Geographic Reference Point

GRVTY – Gravity

H

H - Hour, Hot

HCU - Hydraulic Control Unit

HDG/S - Heading Selected

HDL - Handle

HLD - Hold

HMU - Hydro Mechanical Unit

HPV - High Pressure Valve

I

IDG - Integrated Drive Generator

IGN - Ignition

IMM - Immediate

INB - Inbound

INBO - Inboard

INCREM - Increment

INIT - Initialization

INR - Inner

INTCP - Intercept

I/O - Input/Output

I/P - Input or Intercept Profile

IP - Intermediate Pressure

IPC - Intermediate Pressure Checkvalve

IPPU - Intermediate Position Pick-off Unit

ISIS - Integrated Standby Instrument System

ISOL – Isolation

J

K

L

LAF - Load Alleviation Function

LAT - Latitude

LAT REV - Lateral Revision

LCN - Load Classification Number

L/G - Landing Gear

LGCIU - Landing Gear Control Interface Unit

LGPIU - Landing Gear Position Indicator Unit

LIS - Localizer Internal Smoothing

LK - Lock

LL - Latitude/Longitude

LLS - Left Line Select Key

LNAV - Lateral Navigation

LONG - Longitude

LRU - Line Replaceable Unit

LSK - Line Select Key

LVL - Level

LVL/CH - Level Change

LW - Landing Weight

M

M - Magenta, Mach, Meter

MAG DEC - Magnetic Declination

MAG VAR - Magnetic Variation

MAX CLB - Maximum Climb

MAX DES - Maximum Descent

MAX END - Maximum Endurance

MCDU - Multipurpose Control and Display Unit

MCU - Modular Concept Unit

MDA - Minimum Descent Altitude

MECH - Mechanic

MFA - Memorized Fault Annunciator

MLS - Microwave Landing System

MMR - Multi-Mode Receiver

MN - Mach Number

MRIU - Maintenance and Recording Interface Unit

MSA - Minimum Safe Altitude

MSU - Mode Selector Unit

N

N - Normal, North

NAVAID - Navigation Aid (VOR/DME)

ND - Navigation Display

NW - Nose Wheel

O

OBRM - On Board Replaceable Module

OFF/R - Off Reset

OFST - Offset

O/P - Output

OPP - Opposite

OPT - Optimum

OUTB - Outboard

OUTR - Outer

OVBD - Overboard

OVSPD – Overspeed

P

P-ALT - Profile Altitude

PB - Push Button

PBD - Place/Bearing/Distance Waypoint

PBX - Place-Bearing/Place-Bearing Waypoint

PC - Pack Controller

P-CLB - Profile Climb

P-DES - Profile Descent

PDU - Pilot Display Unit

PFD - Primary Flight Display

PHC - Probe Heat Computer

P-MACH - Profile Mach

POB - Pressure Off Brake

PPOS - Present Position

P-SPEED - Profile Speed

PPU - Position Pick-off Unit

PR - Pressure

PRED - Prediction

PROC - Procedure

PROC T - Procedure Turn

PROF - Profile

PROTEC - Protection

PRT - Printer

PT - Point

PTU - Power Transfer Unit

Q

QRH - Quick Reference Handbook

QT – Quart

R

R - Right, Red

RACC - Rotor Active Clearance Control

RAT - Ram Air Turbine

RCDR - Recorder

RCH - Small unit of measurement

RCL - Recall

RCVR - Receiver

R/I - Radio/Inertial

RLSK - Right Line Select Key

RMP - Radio Management Panel

RNG - Range

RNP - Required Navigational Performance

RPTG - Repeating

RQRD - Required

RSV - Reserves

RTOW - Regulatory TakeOff Weight

S

S - Slat Retraction Speed, South

SC - Single Chime

S/C - Step Climb

SD - System Display

SEL - Selector

STAT INV - Static Inverter

S/D - Step Descent

SDAC - System Data Acquisition Concentrator

SDCU - Smoke Detection Control Unit

SEC - Spoiler Elevator Computer

SFCC - Slat Flap Control Computer

SLT - Slat

SPD LIM - Speed Limit

SPLR - Spoiler

SRS - Speed Reference System

STEER - Steering

STS - Status

SW - Switch

SWTG - Switching

SYNC – Synchronize

T

T - Temperature

TGT - Target

THR - Thrust

THS - Trimmable Horizontal Stabilizer

TK - Tank, Track Angle

TKE - Track Angle Error

TMR - Timer

TLA - Thrust Lever Angle

TOGW - Takeoff Gross Weight

TOW - Takeoff Weight

T-P - Turn Point

T-R - Transmitter-Receiver

TROPO - Tropopause

TRU - Transformer Rectifier Unit

TTG - Time To Go

U

UASS - Unofficial Airbus Study Site

UFD - Unit Fault Data

ULB - Underwater Locator Beacon

UNLK - Unlock

UTC - Universal Coordinated Time

V

VBV - Variable Bypass Valve

VC - Calibrated Airspeed

V/DEV - Vertical Deviation

VEL - Velocity

VFE - Max Flaps Extended Speed

VFEN - VFE Next

VM - Maneuvering Speed

VMIN - Minimum Operating Speed

VNAV - Vertical Navigation

VOR-D - VOR-DME

VSC - Vacuum System Controller

VSV - Variable Stator Vane

W

W - White, West, Weight

WAI - Wing Anti-Ice

WBC - Weight and Balance Computer

WHC - Window Heat Computer

WTB - Wing Tip Brake

WXR - Weather Radar

X

XCVR - Transceiver

XFR - Transfer

Y

Y – Yellow

Z

ZC - Zone Controller

ZFCG - Zero Fuel Center of Gravity

Now that you've memorized the above terms — tell me what they actually mean? This is the technical side of modern day airliner flying.

If you grew up with an iPad it's likely much easier to absorb this new technology compared to those of us cave dwellers who thought the Dewy Decimal System was advanced Calculus.

In reality, becoming a Captain is a reward for patiently putting up with the lengthy time spent watching and learning from the right seat.

Any person who shows up prepared and on time for trips and check rides (while avoiding 'little chats' in the Chief Pilots office) has the makings of a great airline captain. Add to it the following traits and you have a good idea of the ideal captains job description.

An airline captain must possess the ability to:

-   Absorb the technical knowledge necessary to have a working knowledge of the many complex systems of a modern day airliner to include, hydraulic, pneumatic, electrical, flight controls, environmental and flight management systems.

-   Demonstrate recovery and emergency maneuvers to a satisfactory level to include, handling engine failures on takeoff, engine fire in flight, electrical failures and fire, hydraulic system failures, flight control failures, wind shear recovery, pressurization failure — and the list goes on and on.

Author's Note

There are virtually thousands of things that can go wrong on a modern airline, normally only those emergencies that have historically reoccurred and those most critical to safety of flight are emphasized during training. Most airplane mechanical's are just that — mechanicals — kind of like a flat tire on an auto. They're routine in nature and do not pose a danger to crew or passengers. True emergencies, that could be considered life threatening, are extremely rare.

- Have the ability and strength of character to command and respect your crew while managing cockpit resources effectively. Communication skills while in stressful situations are without exception a necessity. During an emergency getting a case of traumatic lock-jaw (inability to speak) could have devastating consequences.

- Good health and the ability to sleep on the back side of your circadian clock are traits or habits that all pilots try to acquire (most never do) since duty days are often 12 or more hours, and "Red Eye" flights still exist. Proper rest is one of the most difficult issues aircrews face. This is not an 8-5 job!

Until now, I've deliberately avoided the typical HR job description of an airline pilots responsibilities — choosing to hit the high points as I remember them — so in the interest of fairness to hard working folks at HR departments everywhere here's a fairly accurate accounting of what they might use for a Captains job description.

The following is a slightly modified version of a generic pilot job description as it appears on **www.americasjobexchange.com**. The text has been modified slightly.

"A captain is responsible for flying a plane from one place to another. Greets crew, performs checklist, ensures cargo has been loaded, checks fuel and weather conditions, and performs takeoff and landing procedures.

Primary Responsibilities

- Fly for airlines that transport people and cargo on a fixed schedule.

- Perform pre-flight checklist on engines, hydraulics, and other systems.

- Take guidance information from air traffic controllers.

- Ensure that all cargo has been loaded and that the aircraft weight is properly balanced.

- Start engines.

- Use cockpit instruments.

- Inform passengers if turbulence is worsening.

- Turn on fasten seatbelt sign.

- Communicate with flight attendants and perform crosschecks.

- Receive takeoff and landing instructions.

- Handle flight emergencies.

- Operate all controls and steer aircraft if other pilot is incapacitated.

- Greet people leaving or coming aboard plane.

- Check fuel and weather conditions.

- Maintain flight schedules and alert ground and crew to possible delays.

- Take turns flying the plane to avoid fatigue.

- Review flight plan with dispatch which details the altitude for the flight, route to be taken and amount of fuel required.

- Regularly check aircraft's performance and position.

- React to environmental changes and altitudes.

- Update aircraft logbook.

- Note any incidents that occurred during the flight.

- Ensure noise regulations are followed during take off and landing."

Sounds simple? — think again!

Now that you know what the job entails, take a look at what it takes to get a foot in the door of a typical Legacy Air Carrier.

Generally the basic job requirements for most major airlines are similar to those shown below:

- At least 23 years of age. (Maximum ages are not usually listed for EEOC purposes.)

- Graduate of a four-year degree program from a college or university accredited by a U.S. Dept. of Education recognized accrediting organization. Degrees obtained from a non-U.S. institution must be evaluated for equivalency to U.S. degrees by a member organization of the National Association of Credential Evaluation Services (NACES). (On rare occasions individuals without a college degree, but with excellent flying credentials and experience are considered.)

- Current passport or other travel documents enabling bearer to freely exit and re-enter the U.S. (multiple reentry status) and you must be legally eligible to work in the U.S. (possess proper working documents).

FAA Requirements

- FAA commercial fixed-wing pilot license with instrument and multi-engine rating.

- Current FAA First Class Medical Certificate.

- Meets minimum FAA ATP requirements including successful completion of ATP written. (ATP – Airline Transport Pilot rating — highest rating possible.)

Flight Time Requirements

- Minimum of 1,500 hours of total documented flight time.

- Minimum of 1,000 hours of fixed wing turboprop or turbofan time. When evaluating flight time of applicants meeting basic qualifications, consideration will be given to, among other things, quality, quantity, rec ency, and verifiability of training; complexity of aircraft flown; types of flight operations; and hours flown as PIC in turbine powered aircraft.

Other Requirements

- Applicants invited to interview must provide appropriate documentation [to support] all flight hours and college transcripts.

- FCC Radiotelephone Operator's Permit.

- You will be required to take and pass a DOT required pre-employment drug test.

- You will undergo a TSA required fingerprint based Criminal History Records Check and company background check.

All applicants are normally required to meet the above criteria just to get a foot in the door of most major airlines. Usually, accepted applicants are better educated, have more flight hours, and are as polished as a candidate running for a congressional seat.

I've flown with airline pilots who have doctorate degrees, and some were doctors, attorneys or dentist before becoming airline pilots. This is a challenging job that attracts very accomplished and motivated individuals.

Once HR reviews your application, and **if you are among the top qualifiers**, you'll likely be extended an interview invitation when hiring begins.

The major airline pilot interview is one of the most intense reviews of one's character, credentials, and flying skills imaginable. This is your dream job. One botched answer to an interview question, one addition error on you log book flight hours or worse yet; a poorly flown simulator ride, or a tarnished background investigation and your dream is shattered — sometimes forever.

Note: If you have a DWI or have been arrested for any reason it will most likely show-up in your background investigation. Having a good answer for these indiscretions will probably not be enough — airlines are not in the habit of hiring [or interviewing] pilot's that might, through an error in judgment, place them in a libelous position.

Most airlines now require a flying skills test in a full motion simulator. Major's usually put you in one of their wide-body airplane simulators if available. This can be intimidating if the largest thing you've ever flown is a light twin engine aircraft. Here's a pearl of wisdom for those that might interview with a major. All airplanes fly the same — 'big iron birds' fly just like a light airplane, the most notable difference is the controls feel heavier and it's just slower to react to control inputs. Besides, on these pre-employment 'sim-checks' an instructor pilot will help you with power settings and aircraft configuration changes (flaps, gear, etc.) What they're looking for is adaptability and airmanship skills: can you fly a big plane on instruments and fly an approach to touchdown? They usually throw-in a holding pattern clearance for good measure — this can be a complicated maneuver for an inexperienced pilot. It requires flying to an electronic navigation aid or fix and entering a defined racetrack (holding) pattern. It sounds simple, but it can be daunting and confusing while flying on instruments. Remember, pilots can't stop and re-group or ask directions. Plus all this is happening at speeds over 200 miles per hour.

Human Resource people at most airlines are tasked with a very tall order when screening pilot candidates. Selecting an individual that will eventually be entrusted with a jet costing hundreds of millions of dollars, not to mention the priceless lives of its passengers is not taken lightly. No HR person wants to hire an accident waiting to happen.

As I said earlier a captain is well vetted by the time they sew on that fourth stripe. The 'first look' HR people are only the first of many to observe you during your career.

As technology evolves the airline captain job description will, out of necessity, change to reflect the times. Keeping current with rapidly evolving technology requires constant retraining for everyone in aviation, but most especially, the airline pilot.

CHAPTER 3

The Truth And Nothing But The Truth

Over the many years of my career I've been approached by literally hundreds of people who seemed fascinated with my occupation and curious about the many different aspects of a pilots life. Many of the questions are similar; some humorous, and others technical in nature. The following is a list of the most common questions I remember with my candid answers.

Q - **What is mandatory retirement age for an airline pilot?**

A - Currently the FAA requires all scheduled airline pilots to retire before their 65th birthday. This includes first officers and flight engineers (second officers).

Q - *What's a typical career path for becoming an airline pilot?*

A - **Year 1:** Acquire Private, Instrument, Commercial, and Multi-Engine Flight Instructor Certificates. This will take about 250 hours of flying and will cost around $35,000. Alternatively, with a four-year college degree and in some instances a two-year college degree you can apply for flight training in the military. This will require a six to eight year commitment — but after leaving the service as a pilot you'll normally go to the top of the interview stack at an airline's Human Resources department.

Year 2-3: (If not a military pilot) Work as a flight instructor, log hours while teaching students to fly; acquire about 750 hours of additional flight time earning $5-25 per hour. Low pay, long days, but great experience.

Airline Job 1: With 1,000 hours logged, apply for and hopefully find a regional airline job as a first officer (co-pilot) earning $16,000 - $20,000 the first year. (Typically pay doesn't get much better the second or third year as a co-pilot.)

Airline Job 2: Upgrade to regional airline captain — usually happens around the three year mark in a good hiring climate. You will have obtained your Airline Transport Pilot certificate at your own expense sometime after acquiring the required 1,500 logged flight hours. This will cost you several thousand dollars. You will earn about $45,000 the first year as a regional airline captain, building up at least 1000 hours of Pilot-in-Command time (takes approximately 1.5 years.) Your log books will now reflect over 3,000 – 5,000 hours flight time, and you've invested approximately 6 – 7 years of your life for the ultimate flying job that is still years away. You're now ready for Job 3.

Airline Job 3: Working for a major airline as a first officer earning $40,000 - $50,000 the first year. At a large airline with 10,000 plus pilots on the seniority list, you can spend anywhere from seven to 16 years before having a seniority number high enough to upgrade to captain. On average from the time you log your first flight hour it takes about 20 years to become a captain for a major airline. That's about twice as long as it takes to become a doctor. Please note this time frame is for pilots fortunate enough to fly for existing legacy carriers. For startups and airlines less than 25 years old you might qualify as a captain in half the time from your first flight lesson to wearing captains stripes.

Airline Job 4: Working at a major airline as a captain earning $150,000-$300,000 per year. Once you're a wide body (jumbo jet) captain, you're truly on top-of-the-pyramid professionally. You are the envy of every pilot that looks up as you pass them on the taxiway. You didn't earn the job because of your brilliant flying skills, bravery, or dashing good looks — you're just reaping the reward of patience with a slow climb up a lengthy seniority list.

Age 65: Mandatory retirement from any scheduled airline job. Retire to Florida, buy a small plane and annoy your non-flying golf buddies with mostly made-up stories about your exciting overnights in exotic ports of call.

Q - *Do you need uncorrected 20-20 vision to fly?*

A - No, but it must be correctable to 20-20 with some restrictions on uncorrected visual acuity. Contact a FAA medical examiner for current criteria if interested.

Q - *How much do airline pilots work a month?*

A - Pilots are typically scheduled to fly 75-90 hours a month. That's time measured from pushback at the departing airport until brakes are set, with chocks in at the destination. Both pilots and flight attendants are only paid for actual flying time.

From the time a crewmember checks in for a flight (usually 1 hour before scheduled departure) and all during the preflight and boarding process they are not being paid. Add to this, the following no pay periods can also be part of a flight crewmembers day. Here's what I mean; your duty day (starting with your check-in time at work) can be up to 16 hours long for a two pilot crew. During that time they're only paid for the actual flight time, that can be eight hours or less. Essentially you've been on the job for 16 hours but are only paid for eight hours (or what you've actually flown.)

Looking back over my pay sheets for the last 20 years, I was startled to discover that I worked (on average) a 58-hour workweek to get 20 hours of pay. I commuted most of my life over a thousand miles to and from work. Factoring in the commute, I was away from home about 65 hours a week to earn 20 hours of pay.

The Federal Aviation Administration determines flying hour and duty time limits. Based on my estimate of 58 hours on the job to get 20 hours of pay, and if my math is correct, in order to be paid for 1,000 hours of flight time annually (during my career I usually flew around 900 hours a year) I'd have to be on the job 2,900 hours a year.

A normal non-flying 40-hour workweek yields 2,000 hours a year on the job.

Frankly, working 2,000 hours a year sounds much better than 2,900 hours. A pilot's working life, 30 years on average, requires you to be on the job about 27,000 hours longer than a land lover. Put another way — a pilot spends approximately 1,125 days or 3.08 years longer on the job than a 40-hour per week employee. I loved my pilot career, but obviously didn't consider the extended work time necessary for this type of work. Most pilots don't. That's why we're different.

In the future, new and higher FAA hour limits will further dilute a pilot's real hourly rate. Fortunately pilot pay has increased marginally since the late 90s, that's primarily because pilots are flying more hours in a month.

* * *

For those of you with 40-hour workweeks, envious of your pilot neighbor who always seems to be home, here's the reality; As far as workweek hours are concerned, don't believe what you've read about pilots working 75 hours a month, that's pure fantasy.

Here's an eye opener. My lawn guy makes $65/hr, the air-conditioning guy charges me $110/hr, plumber $115/hr — the person flying your $100 million plane is usually making much less per hour.

A Captain making $200,000 per year flying a theoretical 1,000 hours a year would have earned $68.96 per hour while on the job. A first officer making $100,000 a year would be making $34.58 per hour. (These figures are based on my previous assumptions about time actually on the job.) With an increase in flight time limitations that hourly rate will likely be much lower. What's up with that? I guess it all boils down to career choices. But honestly, pilot pay is the best bargain airline management ever negotiated.

Here's another interesting comparison.

I pay my family doctor $100 for a 15-minute consult. He sees 30 patients a day, five days a week. That's $15,000 a week times 48 weeks a year — he takes off a month for vacation. That's $720,000 a year gross. Even with a 50% overhead that's still a nice $360,000 per year. Not bad for someone who spent half as much time as a legacy airline captain obtaining his end-job.

For the many talented physicians in the world — you've earned your high pay, and you have my deepest respect. I've just used you as a convenient, albeit altruistic example. As for my plumber, air-conditioning tech and lawn guy — I seriously need to re-negotiate their pay rates.

* * *

The following few pages are extracts from a 314-page document by the Federal Aviation Administration. The document describes the new rules for flight crew duty time limitations of part 121 commercial airline pilots.

Regulations 14 CFR Parts 117, 119 and 121 are complex and confusing — but well intended. Before long pilots will need law degrees to interpret flying regulations — its becoming that complex. Being an airline pilot places you under the enforcement umbrella of the federal government. Failure to comply with the monumental amount of regulations governing pilots could result in certificate revocation or even prison time.

The basic rule for flight time limitations is you cannot be scheduled for more than 1,000 hours of flight in a calendar year; no more than 100 hours in a month or 30 hours in a week. (Don't confuse 'flight time' with 'duty time'.) It's far from that simple. I've selected pages that describe the FAA's new rules regarding duty and rest requirements. This might put you to sleep — but hopefully it will make you realize how governed an airline pilots life can be.

Remember, this is an extract from a 90,000 plus word document — most of which pertains to background information in support of the rule changes. It's fascinating to see how our government thinks.

DEPARTMENT OF TRANSPORTATION

Federal Aviation Administration

14 CFR Parts 117, 119, and 121

Docket No.: FAA-2009-1093; Amdt. Nos. 117-1, 119-16, 121-357

RIN 2120–AJ58

Flightcrew Member Duty and Rest Requirements

AGENCY: Federal Aviation Administration (FAA), DOT.

ACTION: Final rule.

SUMMARY: This rule amends the FAA's existing flight, duty and rest regulations applicable to certificate holders and their flightcrew members operating under 14 CFR Part 121. The rule recognizes the universality of factors that lead to fatigue in most individuals and regulates these factors to ensure that flightcrew members in passenger operations do not accumulate dangerous amounts of fatigue. Fatigue threatens aviation safety because it increases the risk of pilot error that could lead to an accident. This risk is heightened in passenger operations because of the additional number of potentially impacted individuals. The new requirements eliminate the current distinctions between domestic, flag and supplemental passenger operations. The rule provides different requirements based on the time of day, whether an individual is acclimated to a new time zone, and the likelihood of being able to sleep under different circumstances.

DATES: Effective [INSERT DATE 2 YEARS AFTER DATE OF PUBLICATION IN THE FEDERAL REGISTER].

The FAA is issuing this final rule to address the risk that fatigue poses to passenger operations conducted under 14 CFR part 121. Part 121 applies to the majority of flights flown by the American public. As such, changes to the existing flight, duty and rest rules in part 121 will directly affect the flying public. This rule applies to all part 121 passenger operations, including traditional scheduled service and large charter operations. The FAA has removed the existing distinctions between domestic, supplemental and flag passenger operations because the factors leading to fatigue are universal and addressing the risk to the flying public should be consistent across the different types of operations.

This final rule addresses fatigue risk in several ways. The underlying philosophy of the rule is that no single element of the rule mitigates the risk of fatigue to an acceptable level; rather, the FAA has adopted a system approach, whereby both the carrier and the pilot accept responsibility for mitigating fatigue. The carrier provides an environment that permits sufficient sleep and recovery periods, and the crewmembers take advantage of that environment. Both parties must meet their respective responsibilities in order to adequately protect the flying public.

The final rule recognizes the natural circadian rhythms experienced by most people that causes them to be naturally more tired at night than during the day. Under the final rule, flightcrew members will be able to work longer hours during the day than during the night. Significant changes in time zones, a situation unique to aviation, are accounted for to reduce the risk to the flying public posed by "jetlag".

The FAA has decided against adopting various provisions proposed in the NPRM. The final rule does not apply to all-cargo operations, although those carriers have the ability to fly under the new rules if they so choose. The proposal that carriers meet certain schedule reliability requirements has been dropped, as has the proposed requirement that carriers evaluate flightcrew members for fatigue. The FAA has determined that these provisions were either overly costly or impractical to implement.

1. *Fitness for Duty.*

This rule places a joint responsibility on the certificate holder and each flightcrew member. In order for the flightcrew member to report for an FDP properly rested, the certificate holder must provide the flightcrew member with a meaningful rest opportunity that will allow the flightcrew member to get the proper amount of sleep. Likewise, the flightcrew member bears the responsibility of actually sleeping during the rest opportunity provided by the certificate holder instead of using that time to do other things. The consequence of a flightcrew member reporting for duty without being properly rested is that he or she is prohibited from beginning or continuing an FDP until he or she is properly rested.

2. *Fatigue Education and Training.*

Part 121 air carriers are currently statutorily-required to annually provide, as part of their Fatigue Risk Management Plan, fatigue-related education and training to increase the trainees' awareness of: (1) fatigue; (2) "the effects of fatigue on pilots;" and (3) "fatigue countermeasures." Today's rule adopts the same standard of training as required by the statute. In addition, today's rule adopts a mandatory update of the carriers' education and training program every two years, as part of the update to their FRMP. Both of these regulatory provisions merely place the existing statutory requirements in the new flight and duty regulations for the ease and convenience of the regulated parties and the FAA.

3. *Fatigue Risk Management System.*

The FAA proposed a Fatigue Risk Management System (FRMS) as an alternative regulatory approach to provide a means of monitoring and mitigating fatigue. Under an FRMS, a certificate holder develops processes that manage and mitigate fatigue and meet an equivalent level of safety. The FAA is adopting that proposal largely as proposed. The FAA has also decided to extend the voluntary FRMS program to all-cargo operations, which are not required to operate under part 117. Under the FRMS provisions that this rule adds to subparts Q, R, and S of part 121, an all-cargo operator that does not wish to operate under part 117 can nevertheless utilize an FRMS as long as it has the pertinent FAA approval.

Asleep yet? The next question should wake you up.

Q - *Do pilots 'fool around' on layovers?*

A -   This is a loaded question. Pilots, like all of us, are human. I have to be candid for a moment; take a group of people (male and female) that work closely together in an exciting, albeit glamorous atmosphere — send them out of town, often thousands of miles from home. It's only natural that relationships are likely to develop. It's not uncommon in the pilot and flight attendant ranks to have been married more than once, twice and even three times.

Remember we spend far more time away from home than a normal 8 to 5 employee. It's likely a pilot or flight attendant will spend nearly half their working life away from home pursuing this career.

By necessity the industry has inadvertently set pilots (and flight attendants) up for having failed relationships. Think about it —they send you off with usually handsome or attractive people of the opposite sex, and pay for your room in a nice hotel. This has "Something's bound to happen," written all over it — it's human nature.

Honestly, I have no idea how many extra marital affairs pilots have on the road. I know it happens, and I know a lot of pilot's whose second or third spouse is a former flight attendant or another pilot — draw your own conclusions. On the flip side many pilots live wonderful monogamous lives with the person of their dreams and loving kids who are used to not having both parents around during celebrated times.

Note: In 35 years of airline flying I only remember having eight Christmas days, three of my birthdays, and four wedding anniversaries at home. The rest of these celebrations were spent with my crewmembers, flying a trip.

Q - *Do pilots drink alcohol while on duty?*

A - Pilots are regulated by law regarding how much time must elapse from 'bottle-to-throttle' (from last drink to the time you enter the cockpit.) Flying intoxicated is dangerous and reckless. Those caught doing so will likely spend or have spent time behind bars. No airline pilot takes this lightly. Do pilots drink a lot? Probably no more than the average person. What separates us from most people is we usually know when it's time to quit. The majority of pilots I know are casual drinkers at best. Some of you may ask why pilots drink at all. I guess the logical answer is it's something to do during long layovers to kill the time.

The unfortunate reality is that occasionally a pilot does become alcohol reliant to the extent that they need help. To address this serious issue the Airline Pilot Association (ALPA) established the Human Intervention Motivation Study program (HIMS). This rehabilitation program has returned over 4,000 pilots to the cockpit since it began over 40 years ago.

HIMS was developed for airline pilots by airline pilots because a thoughtful, frank physician stood before the 1972 ALPA Board of Directors and opened the taboo subject of alcoholism in the cockpit. At that time no reliable or system-wide method existed to provide help for airline pilots who had alcohol problems.

In the early 70s, pilots diagnosed with alcohol dependency were simply fired from their jobs, lost their FAA medical certification, and were banned forever from the airline piloting profession. The notion that alcoholism was a legitimate disease was alien to most people in the air transport industry; the prevailing attitude was "once a drunk, always a drunk," and as a result, alcohol-addicted pilots stayed in the closet. "Too often, the only way we would discover an alcoholic pilot was when withdrawal seizures struck while on duty," said an FAA official.

Note: The HIMS program has since grown to include dealing with chemical dependency issues among airline pilots. Some airline pilots who are military reservist returning from recent military flying in combat theaters have been legally prescribed "Go Pills," Dexedrine (an amphetamine,) by battlefield physicians to ward off the sleepiness brought on by continuous high stress-long hour missions. The drug has caused addiction problems for some of these pilots. Additionally during down times it was found that "No-Go-Pills," like Ambien were needed for combat weary pilots to sleep. These deserving airline/military aviators have been given a job-saving chance to continue their airline careers thanks to the HIMS program.

You might be asking yourself: is there a chance that your pilots are drunks or drug addicts? The simple answer is no. Because of the responsibility entrusted to airline pilots, their own personal discipline, and availability of programs such as HIMS, (not to mention the guaranteed threat of criminal prosecution and prison time) the likelihood of your crew being inebriated or stoned is virtually impossible. Those pilots who were closeted have jumped at the opportunity to get help, especially if it meant keeping their job after successfully completing the program.

Pilots and flight attendants probably have more random drug test than any profession on earth. You get caught; you're fired. You fess up and if your airline endorses the HIMS or similar programs you might be able to keep your job after a lengthy rehabilitation. Pilots are a responsible group, probably more so than in most every other occupation. We are however, human!

I have spoken to many pilots who've gone through the HIMS program. Their worst habit now seems to be chewing gum. It's a testament to ALPA's conviction toward protecting its pilot membership that these responsible aviators are now given a second chance. The pilots entering this program have a sincere desire to do the right thing. This speaks volumes about their character.

Q - What's the most dangerous thing an airline pilot does?

A - Drive to the airport — it's far more dangerous than flying.

Q - *How fast can an airliner fly?*

A - It depends on the type of airplane. All commercial jets are structurally limited to a maximum indicated airspeed typically shown as a Never Exceed Speed in knots at lower altitudes, and referred to as a Maximum Mach Operating number at higher altitudes where the air is to thin to measure numerical speeds accurately.

At altitude, the worlds largest airliner, the Airbus A-380, has a limiting Mach number of .88 Mach — translated it can fly at 88 percent of the speed of sound at its max service ceiling of 43,000 feet. That's 669,862,205 mph in case you're wondering.

Q - *Why does a pilot walk around the plane before each flight? Isn't that the mechanics job?*

A - The FAA requires that a cockpit crewmember 'walk-around' their aircraft prior to each and every flight. Not all stations have mechanics sitting around waiting to look at your airplane

— so a pilot's exterior preflight bridges the inspection gap. Arguably mechanics can usually give you a better preflight. This is why an FAA certified Airframe and Power plant mechanics must accomplish a complete aircraft preflight at least once each day the aircraft is in service.

Pilots performing an exterior preflight look for the obvious things on their inspections. Fuel and oil leaks, airframe and tire condition, and anything out of the ordinary. It's a good system that has saved airlines from many potentially embarrassing moments. Plus it gives you a chance to stretch your legs and listen to the roar of jet engines. I really do love the smell of jet fuel in the morning.

Q - *How fast do you have to go to takeoff?*

A - It depends on the plane and its gross takeoff weight. Typically most airliners are airborne between 130 to 160 mph. The rotate speed that puts the airplane in a takeoff attitude is carefully calculated before each and every takeoff. When a pilot reaches that speed (known as Vr or Rotate speed) a steady pull back on the control wheel or side stick is applied until the aircraft's pitch reaches take-off attitude. At about the time the optimum pitch attitude is obtained the plane is now at V2 speed. If an engine fails at this point V2 speed is maintained until a safe altitude is reached. At that point the airplane would most likely be reconfigured for an emergency landing.

Q - *Are women pilots better than their male counterparts?*

A - This is usually a question asked by insecure men. To answer properly let me offer a little background. Like male pilots, female aviators are both strong and weak. I have flown with and instructed many female pilots in big jets. Many come from general aviation backgrounds and lack the experience of some of their female and male counterparts who flew heavy transports or fighters in the military. This disparity is neutralized over time. When they finally check out as a captain they will have accumulated thousands of hours as a first officer and the playing field will have long been equalized. Collectively I have found female pilots to be some of the more conscientious aviators I've ever had the pleasure of flying with. They are not better or worse than their male counterparts. Most are a joy to fly with and are highly motivated to succeed in this profession. If you look up front in a big jet and see a woman in the left seat — she's likely an exceptional pilot. The road to being a female airline captain wasn't always easy in this historically male dominated occupation. Female pilots at major airlines didn't really start to appear until the 70s. As of July 2014 approximately 5.12 percent of the commercial airline pilots are female. Fortunately their number is growing.

Q - *I've saved the best and most frequently asked question for last. More times than I can remember, people find out I was an airline pilot and the first question out of their mouth is — "Have you ever seen a UFO?"*

A - This is a difficult question to answer. Not so much because of my intended response, but depending on my testimony I could be labeled a nutjob or worse yet a conspiracy theorist. My friends tell me I'm neither — so here it goes.

I've been a licensed pilot for almost 50 years. Of the 196 countries on earth (including Taiwan) I have flown over all but 16 of them and landed in most. During that time I only remember two occasions (both at night) where I observed (along with the other pilot) lights in the sky that had nonlinear – random like motion. On both occasions we contacted air traffic control and asked if they had radar contact with the traffic we were observing high above us (thinking it might be extremely high altitude secret military operations.) One encounter was a single target, the other about a half dozen objects that were in a loose formation. Neither encounter was picked up by ATC. Were they UFOs, secret military test, or just aberrations? I honestly don't know. On both occasions myself and the other pilot watched these unidentifiable objects for several minutes. By the way, both encounters were over United States soil. Perhaps that's a hint? Regardless, I guess my truthful answer would have to be, yes I've seen flying objects that I couldn't identify. I have not seen little green men, been abducted or followed by Men In Black after these events. This was just a late night observation of extremely fast, randomly moving bright objects in front of my plane. That's the whole truth, and nothing but the truth.

CHAPTER 4

From The Ground To 10,000 Feet

Once passenger boarding is complete and the cockpit door closed, things get very busy for airline pilots. The captain now becomes the ringleader for a show that can have many endings. If his or her directions are spot-on, the flight to your destination will be safe and flawless. If an error is made the consequences can sometimes be disastrous.

In 1981 the FAA imposed the Sterile Cockpit Rule as a result of several crashes where the crews were distracted from their duties by non-essential conversations and activities during critical parts of the flight.

The following is text from that regulation:

U.S. FAR 121.542/135.100, "Flight Crewmember Duties":

- No certificate holder shall require, nor may any flight crewmember perform, any duties during a critical phase of flight except those duties required for the safe operation of the aircraft. Duties such as company required calls made for such non-safety related purposes as ordering galley supplies and confirming passenger connections, announcements made to passengers promoting the air carrier or pointing out sights of interest, and filling out company payroll and related records are not required for the safe operation of the aircraft.

- No flight crewmember may engage in, nor may any pilot in command [captain] permit, any activity during a critical phase of flight that could distract any flight crewmember from the performance of his or her duties or which could interfere in any way with the proper conduct of those duties. Activities such as eating meals, engaging in nonessential conversations within the cockpit and nonessential communications between the cabin and cockpit crews, and reading publications not related to the proper conduct of the flight are not required for the safe operation of the aircraft.

- For the purposes of this section, critical phases of flight includes all ground operations involving taxi, takeoff and landing, and all other flight operations conducted below 10,000 feet, except cruise flight.

Note: Taxi is defined as "movement of an airplane under its own power on the surface of an airport".

I started my airline career in 1973 as a flight engineer on a Lockheed L-188 Electra, a big four-engine turboprop. I wish this rule had been in effect back then. Distractions below 10,000 feet were many and mostly nonessential. Most of the time I spent during climb-out was watching the aircraft systems, getting request from the flight attendants, and making calls to the company with our out-off and ETA times. I hardly ever had a chance to look out the window for traffic. Jump ahead 40 years and even though flight attendants don't bother you below 10,000 feet unless it's an emergency, and out, off, and

ETA times are automatically sent electronically, there's still a great deal of distractions.

Here's a short but by no means complete list of all a captain usually deals with once the cockpit is sterile and preparation is made for engine start and takeoff − the first officer is equally as busy.

- Verify ATC clearance.

- Direct F/O to call for pushback clearance.

- Contact the pushback crew while monitoring ramp control frequency for changes in your 'push' clearance.

- When cleared by pushback crew to start engines call for engine start checklist.

- Command and monitor engine start while assuring pushback crew is following directions given by ramp control.

- Once all engines are started and pushback crew has requested "Brakes set − are we cleared to disconnect," the captain sets the brakes and clears 'push' crew to disconnect from the aircraft.

- Captain calls for "Before Taxi Check."

- After before taxi check is complete Captain request F/O obtain taxi clearance.

- After an often-complex taxi routing is issued by ground control both Captain and First Officer confirm route by looking at a taxiway diagram. The airplane finally begins to move under its own power. Taxiways are named by alphabet or alpha-numerically. Airports like Chicago O'Hare have such a complex taxiway system that even the most seasoned captains can become disoriented.

- The Captain can't see his wing tips or tail so they taxi slowly through congested ramp areas and complex taxiways to avoid colliding with another aircraft or obstacle. This is harder than it sounds.

- A Boeing 777 has a wingspan of almost 200 feet and a length of just over 242 feet. Depth perception only helps so much when taxing these large jets. Following the painted taxiway centerlines are really the only reliable and visible means of assuring you at least remain on the pavement. Taxiing becomes even more challenging when the captain is responding to checklist while looking at and confirming switch positions — it's like texting and driving. It's legal in an airplane, but a bit unnerving.

- Approaching the number one position for takeoff the Captain calls for the, "Before takeoff check," as soon as their cleared onto the runway.

- Once cleared for takeoff the pilot flying assumes control of the aircraft (legs of a trip are usually swapped between Captain and First Officer to assure both pilots stay proficient.)

- The Captain is the only one that can abort the takeoff roll prior to decision speed — called V1. In the event of an abort, things happen lightning fast. If the tower is not paying attention someone could land on top of you after an abort, and because of heat build up in the tires during a max effort stop — the tires could explode or hot brakes could catch the airplane on fire. If that happens you might find yourself on an escape slide ride to the outside world — the Captain orchestrates everything that happens after an abort. A wrong call and passengers and crew might be injured for no reason.

- Assuming your takeoff is successful, you now climb out in the most congested airspace imaginable.

- After the airplane is cleaned up (gear and flaps retracted) and you're safely out of the terminal area the airplane is accelerated to 250 knots and maintains that speed until passing 10,000 feet. Even with a traffic collision avoidance system (TCAS) installed (all airliners have them) both pilots keep their heads on a constant swivel to avoid hitting other aircraft. The fancy avoidance systems help greatly, but are not fool proof, especially around busy airport areas.

- Above 10,000 feet the captain turns off the sterile cockpit light in the main cabin, alerting flight attendants that they can now call or access the cockpit.

- Once you arrive at cruise altitude the cockpit crew continues to navigate, monitor fuel consumption, and aircraft systems while talking to a number of air traffic controllers along their route. To keep ahead of the plane they can get hourly updates on the destination weather. If the planned arrival airport has weather that precludes an approach and landing — well, now things get interesting and their passengers are about to have a bad and inconvenient day.

As serious as flying is there always seems to be time for levity. Both controllers and pilots have potentially stressful jobs. Here's a look at some stress relieving exchanges between pilots and controllers. These stories have been around for years, and while I can't attest to their authenticity or authorship I have heard similar exchanges many times during my career. Enjoy the humor!

Tower: "Delta 351, you have traffic at 10 o'clock, 6 miles!"

Delta 351: "Give us another hint! We have digital watches!"

* * *

Center: "TWA 2341, for noise abatement turn right 45 degrees."

TWA 2341: "Center, we are at 35,000 feet. How much noise can we make up here?"

Center: "Sir, have you ever heard the noise a 747 makes when it hits a 727?"

* * *

From an unknown aircraft waiting in a very long takeoff line: "I'm f...ing bored!"

Ground Traffic Control: "Last aircraft transmitting, identify yourself immediately!"

Unknown aircraft: "I said I was f...ing bored, not f...ing stupid!"

* * *

O'Hare Approach Control to a 747: "United 329 heavy, your traffic is a Fokker, one o'clock, three miles, Eastbound."

United 329: "Approach, I've always wanted to say this... I've got the little Fokker in sight."

* * *

A student became lost during a solo cross-country flight. While attempting to locate the aircraft on radar, ATC asked, "What was your last known position?"

Student: "When I was number one for takeoff."

* * *

A DC-10 had come in a little hot and thus had an exceedingly long roll out after touching down.

San Jose Tower Noted: "American 751, make a hard right turn at the end of the runway, if you are able. If you are not able, take the Guadeloupe exit off Highway 101, make a right at the lights and return to the airport."

* * *

A Pan Am 727 flight, waiting for start clearance in Munich, overheard the following:

Lufthansa (in German): "Ground, what is our start clearance time?"

Ground (in English): "If you want an answer you must speak in English."

Lufthansa (in English): "I am a German, flying a German airplane, in Germany. Why must I speak English?"

Unknown voice from another plane (in a beautiful British accent): "Because you lost the bloody war!"

* * *

Tower: "Eastern 702, cleared for takeoff, contact Departure on frequency 124.7"

Eastern 702: "Tower, Eastern 702 switching to Departure. By the way, after we lifted off we saw some kind of dead animal on the far end of the runway."

Tower: "Continental 635, cleared for takeoff behind Eastern 702, contact Departure on frequency 124.7. Did you copy that report from Eastern 702?"

Continental 635: "Continental 635, cleared for takeoff, roger; and yes, we copied Eastern... we've already notified our caterers."

* * *

One day the pilot of a Cherokee 180 was told by the tower to hold short of the active runway while a DC-8 landed. The DC-8 landed, rolled out, turned around, and taxied back past the Cherokee. Some quick-witted comedian in the DC-8 crew got on the radio and said,

"What a cute little plane. Did you make it all by yourself?"

The Cherokee pilot, not about to let the insult go by, came back with a real zinger: "I made it out of DC-8 parts. Another landing like yours and I'll have enough parts for another one."

* * *

The German air controllers at Frankfurt Airport are renowned as a short-tempered lot. They not only expect you to know your gate parking location, but how to get there without any assistance from them. This is an amusing exchange between Frankfurt ground control and a British Airways 747, call sign: Speedbird 206.

Speedbird 206: " Frankfurt , Speedbird 206 Clear of active runway."

Ground: "Speedbird 206. Taxi to gate Alpha One-Seven."

The BA 747 pulled onto the main taxiway and slowed to a stop.

Ground: "Speedbird, do you not know where you are going?"

Speedbird 206: "Stand by, Ground, I'm looking up our gate location now."

Ground (with quite arrogant impatience): "Speedbird 206, have you not been to Frankfurt before?"

Speedbird 206 (coolly): "Yes, twice in 1944, but it was dark, -- And I didn't land."

* * *

While taxiing at London's Gatwick Airport, the crew of a US Air flight departing for Ft. Lauderdale made a wrong turn and came nose to nose with a United 727.

An irate female ground controller lashed out at the US Air crew, screaming: "US Air 2771, where the hell are you going? I told you to turn right onto Charlie taxiway! You turned right on Delta! Stop right there. I know it's difficult for you to tell the difference between C and D, but get it right!"

Continuing her rage to the embarrassed crew, she was now shouting hysterically: "God! Now you've screwed everything up! It'll take forever to sort this out! You stay right there and don't move till I tell you to! You can expect progressive taxi instructions in about half an hour, and I want you to go exactly where I tell you, when I tell you, and how I tell you! You got that, US Air 2771?"

"Yes, ma'am," *the humbled crew responded.*

Naturally, the ground control communications frequency fell terribly silent after the verbal bashing of US Air 2771. Nobody wanted to chance engaging the irate ground controller in her current state of mind. Tension in every cockpit out around Gatwick was definitely running high. Just then an unknown pilot broke the silence and keyed his microphone, asking:

"Wasn't I married to you once?"

* * *

My thanks to the many nameless pilots who've perpetuated these stories. Your quick wit has helped many fellow airline pilots through the long, long days in the cockpit.

Humor withstanding, airspace globally is becoming more crowded despite safe distances between aircraft being reduced to almost impractical limits. This has been done to allow more planes to occupy a block of airspace – kind of an oddball solution to crowded skies. Add to that the near term introduction of large unmanned drones flying in the national airspace system and collision avoidance will take on a whole new meaning.

On a side note I've written two novels about drones. I have a broad understanding of the subject and offer the following:

Unmanned aircraft are here to stay — unmanned does not mean they are un-piloted. There will, at least for the immediate future, always be a need for some person/pilot to initiate guidance of unmanned aerial vehicles. The distant future may relegate all commercial airline pilots to ground based cockpits. When that happens, these people will still have a captain's responsibility, but their skill sets will be even more demanding.

I can only imagine what an airline pilot's life will be like a hundred years from now. Sterile airspace will likely exist from the ground based control station all the way up to the edge of space. It's not as far fetched as you might think. The Air Force RQ-4 Global Hawk drone known militarily as an Unmanned Aerial System or UAS is 47.6 feet long with a wingspan of 130.9 feet. It has a range of 12,300 nautical miles. It's as big as a regional airline jet and can be controlled — from the ground — by a drone pilot sitting in a trailer half way around the world. For my under 12 readers you might want to take up model drone flying as a hobby. It's cheaper than flying lessons.

CHAPTER 5

The Airline Pilot's Oath

For over a quarter of a century I have been a proud member of ALPA — The Air Line Pilots Association, International. It is the largest airline pilot union in the world representing over 51,000 pilots. Founded in 1931, the Association is chartered by the AFL-CIO and the Canadian Labor Congress. Known internationally as US-ALPA, it is a member of the International Federation of Air Line Pilots Associations.

Before reading ALPA's Code of Ethics it's important to understand that, unlike the Hippocratic Oath where physicians swear to preserve life, ALPA's Code of Ethics sets a standard of excellence that protects human life — not just one life on an operating table, but hundreds, potentially thousands of lives each day we pilot an airliner.

For most ALPA pilots this code is a not-so-subtle reminder of their responsibilities. Before reading the Code of Ethics you need to understand these lofty standards have not always been easy to honor. Putting ethical values ahead of emotional reactions has been tested over way-too-many-airline-strikes and negotiation periods.

Despite the oftentimes rocky road to the "Left seat," pilots seem to have an unwavering determination to "Hang-in-there." Consider the following if you doubt their resiliency.

The percentage of profit a successful major airline claims, which is a surprisingly low number, but significant dollar amount, determines workplace satisfaction. When airlines are profitable employees are usually content. "Keep making money and all is well," is the mantra of most airline executives.

As would be expected from any management team, watching the bottom line has become an all-consuming job. The largest airline expense other than fuel and equipment is undeniably pilot salaries. Attacking pilot earnings is always an easy target when looking at improving bottom lines.

It's true, some pilots make in excess of $250,000 per year. On the opposite end of pilot pay scales a very large percentage are compensated at subpar wages. ALPA fights daily to change this disparity.

Oddly enough historical habits have created this extreme variance in incomes among pilots. Almost from the beginnings of commercial aviation the first pilot hired was usually paid the most — that's what seniority really meant.

Over the years pay gaps between the most senior pilot and most junior have widened. Keep in mind — almost all of these individuals are by FAA standards equally qualified as pilots. Unfortunately seniority-based pay is an almost impossible system to break. The biggest problem would be establishing pay levels satisfactory to senior captains who have come up through the seniority ranks and have already endured the low entry-level pay established by these historical compensation practices. Now that they're making the "Big Bucks" they really don't want to divide what I'll call the salary pie (money budgeted by the airline for pilot pay) and take a pay cut to bring up junior pilot compensation. Regrettably few of those senior pilots probably spent very long at the low pay rate. As legacy carriers rapidly grew early on, hiring numbers most likely soared making the trip to the left seat a much shorter journey. Example: If you're number 10,000 on a seniority list with that number of pilots, you need to wait for 5,000 pilots above you to check out before it's your turn to become a captain (assuming there are 5,000 captains and 5,000 first officers.) That could mean decades before making the big bucks.

Unfortunately upgrades and pay increases are not linear or guaranteed. Typically most airline pay scales peak out at 10-12 years for any given position (Capt., F/O or Engineer.) Once you've reached that anniversary the only way you get a pay raise is by moving to a larger piece of equipment or upgrading. It's a system that doesn't reward those who go above and beyond or have astronaut skill sets. It's strictly a numbers and seniority game — there's no such thing as a merit raise.

By what some would call a faulty design the typical pilot pay scale system usually ends up costing airlines millions in extra training cost. Since moving to a larger piece of equipment usually means a pay raise, pilots are continually upgrading and changing equipment to get the larger salary. These changes in position can happen as often as every two years. The logical system would be to have the same pay rate for respective Captains, First Officers, and Second Officers (Flight Engineers) regardless of equipment.

This could and has worked at larger air carriers. At the very least it cut pilot training cost dramatically.

Are pilots worth more than they're paid? I've broached this subject previously, but I'd like to offer the following:

A large number of pilot new-hires come from the ranks of military aviators. Training of that individual in most cases cost our government over a million dollars. Many of these pilots have flown almost to the edge of space and some have actually orbited Earth. Even these astronauts start at the bottom of the airline seniority list.

It's a fact of life. If you want to fly big jets for a living with hopes of a six-figure retirement income you need to literally pay your dues, regardless of your qualifications when hired. It's not uncommon for junior captains at smaller carriers to have less experience than their military trained First or Second Officers. Over time that usually becomes a meaningless point, especially at the majors where, with rare exception, the person in the captain's seat is usually the most qualified pilot.

What motivates a person to work ungodly hours for low pay leaving their families for days on end? Passion is the only answer I can come up with.

Passion or not, pilot employment can lead to a decent life and respectable wages. During pilot contract negotiations an annual compensation budget is negotiated with management — how it's divided is usually left to the pilot negotiators. The lion's share of this pilot salary budget always goes to those who have been on the property the longest. Its not greed motivated. Really...

All major airline pilots look to their future captainship as being their reward for patience and mediocre pay. For some, rapid advancement due to large growth spurts or mergers puts them in the big buck range sooner than later. For others, the picture is not quite as bright.

It's anyone's guess as to how an airline will do in the future. Giants like Pan Am, National Airlines, Eastern Airlines, TWA, etc., have vanished while some start-ups have flourished. If you're lucky you might become 'captain big bucks' in as little as three years. Usually it takes much, much longer especially for larger airlines with thousands on the seniority list.

In an ebbing economy, few retirements and downsizing can stretch the upgrade time to 18 years and beyond. Some unfortunate few never upgrade to captain and can reach the mandatory retirement age without ever having the seniority to occupy the captain's seat. This is rare, but I've seen it happen.

* * *

It seems strange to me that most people universally accept and expect hospital and physician expenses to run in the hundreds of thousands of dollars to save one person's life. Conversely the crew flying your $100 million, 500 mph, 9-mile-high flying jet, entrusted with your life and the lives of several hundred fellow passengers are viewed by many airline executives as undeserving of compensation greater than a city bus driver. I don't know about you, but I want any person responsible for my life, be it pilot or doctor, to be well compensated and without financial worry that might distract them from their job. Remember being a pilot is one of the most dangerous jobs in the world. I'm not sure doctors or airline executives are even listed on those types of list. Hubris aside, there are a growing number of airline managers who value their employees' worth and, because of their tenacity in troubled financial times, have kept the cabin doors open and pilots employed.

Compensation and work rules are contentious enti-ties pilots and management constantly battle over. We would all like to have a respectable income and a job that satisfies our emotional health and en-trepreneurial spirit. An airline captain's job, despite the initial low wages, often-lengthy apprenticeship and the rigors of being on the road half of your employed life, is in my opinion still the ultimate oc-cupation. The cockpit of a modern jet airliner cruis-ing at 45,000 feet is the most beautiful office in the world. The view is second to none.

By now you should have a small sense of why peo-ple put up with so much to someday occupy an unusually uncomfortable but incredibly expensive seat up front. Individual corporate philosophies aside, the following words — even though geared to the pilot profession — are principles airline pilots truly believe in. Being the person in charge of an airliner's safety can be the loneliest position on or above Earth. Like thousands before me ALPA's Code of Ethics and the integrity found in these words reach beyond the rigors of the profession and have shored me up more than once when making critical decisions. Simply put, reading this code makes me proud to have been an airline Captain.

ALPA Code of Ethics

Author's Note

The following quoted words, although they appear gender specific, apply equally to our many female captains and crew who have fortunately become commonplace in the airline cockpit.

"**An Air Line Pilot** will keep uppermost in his mind that the safety, comfort, and well-being of the passengers who entrust their lives to him are his first and greatest responsibility.

- He will never permit external pressures or personal desires to influence his judgment, nor will he knowingly do anything that could jeopardize flight safety.

- He will remember that an act of omission can be as hazardous as a deliberate act of commission, and he will not neglect any detail that contributes to the safety of his flight, or perform any operation in a negligent or careless manner.

- Consistent with flight safety, he will at all times operate his aircraft in a manner that will contribute to the comfort, peace of mind, and well-being of his passengers, instilling in them trust in him and the airline he represents.

- Once he has discharged his primary responsibility for the safety and comfort of his passengers, he will remember that they depend upon him to do all possible to deliver them to their destination at the scheduled time.

- If disaster should strike, he will take whatever action he deems necessary to protect the lives of his passengers and crew.

An Air Line Pilot will faithfully discharge the duty he owes the airline that employs him and whose salary makes possible his way of life.

- He will do all within his powers to operate his aircraft efficiently and on schedule in a manner that will not cause damage or unnecessary maintenance.

- He will respect the officers, directors, and supervisors of his airline, remembering that respect does not entail subservience.

- He will faithfully obey all lawful directives given by his supervisors, but will insist and, if necessary, refuse to obey any directives

that, in his considered judgment, are not
lawful or will adversely affect flight safety. He
will remember that in the final analysis the
responsibility for safe completion of the flight
rests upon his shoulders.

- He will not knowingly falsify any log or
  record, nor will he condone such action by
  other crewmembers.

- He will remember that a full month's salary
  demands a full and fair month's work. On his
  days off, he will not engage in any occupation
  or activity that will diminish his efficiency or
  bring discredit to his profession.

- He will realize that he represents the airline
  to all who meet him and will at all times
  keep his personal appearance and conduct
  above reproach.

- He will give his airline, its officers, directors,
  and supervisors the full loyalty that is their
  due, and will refrain from speaking ill of
  them. If he feels it necessary to reveal and
  correct conditions that are not conducive to
  safe operations and harmonious relations,
  he will direct his criticism to the proper
  authorities within ALPA.

- He will hold his airline's business secrets in confidence, and will take care that they are not improperly revealed.

An Air Line Pilot will accept the responsibilities as well as the rewards of command and will at all times so conduct himself both on duty and off as to instill and merit the confidence and respect of his crew, his fellow employees, and his associates within the profession.

- He will know and understand the duties of each member of his crew. If in command, he will be firm but fair, explicit yet tolerant of deviations that do not affect the safe and orderly completion of the flight. He will be efficient yet relaxed, so that the duties of the crew may be carried out in a harmonious manner.

- If in command, he will expect efficient performance of each crew member's duties, yet he will overlook small discrepancies and refrain from unnecessary and destructive criticism, so that the crew member will retain his self-respect and cooperative attitude. A frank discussion of minor matters of technique and performance after the flight will create goodwill and a desire to be helpful, whereas sharp criticism and peremptory orders at the moment will result only in the breakdown of morale and an inefficient, halting performance of future duties.

- An Air Line Pilot will remember that his is a profession heavily dependent on training during regular operations and, if in command, will afford his flight crew members every reasonable opportunity, consistent with safety and efficiency, to learn and practice. He will endeavor to instill in his crew a sense of pride and responsibility. In making reports on the work and conduct of his crew members, he will avoid personal prejudices, make his reports factual and his criticisms constructive so that actions taken as a result of his reports will improve the knowledge and skill of his crew members, rather than bring discredit, endanger their livelihood, and threaten their standing in the profession.

- While in command, the Air Line Pilot will be mindful of the welfare of his crew. He will see to it that his crew are properly lodged and cared for, particularly during unusual operating conditions. When cancellations result in deadheading, he will ensure that proper arrangements are made for the transportation of his crew before he takes care of himself.

An Air Line Pilot will conduct his affairs with other members of the profession and with ALPA in such a manner as to bring credit to the profession and ALPA as well as to himself.

- He will not falsely or maliciously injure the professional reputation, prospects, or job security of another pilot, yet if he knows of professional incompetence or conduct detrimental to the profession or to ALPA, he will not shrink from revealing this to the proper authorities within ALPA, so that the weak member may be brought up to the standards demanded, or ALPA and the profession alike may be rid of one unworthy to share its rewards.

- He will conduct his affairs with ALPA and its members in accordance with the rules laid down in the Constitution and By-Laws of ALPA and with the policies and interpretations promulgated therefrom. Whenever possible, he will attend all meetings of ALPA open to him and will take an active part in its activities and in meetings of other groups calculated to improve air safety and the standing of the profession.

- An Air Line Pilot shall refrain from any action whereby, for his personal benefit or gain, he take advantage of the confidence reposed in him by his fellow members. If he is called upon to represent ALPA in any dispute, he will do so to the best of his ability, fairly and fearlessly, relying on the influence and power of ALPA to protect him.

- He will regard himself as a debtor to his profession and ALPA, and will dedicate himself to their advancement. He will cooperate in the upholding of the profession by exchanging information and experience with his fellow pilots and by actively contributing to the work of professional groups and the technical press.

As An Air Line Pilot the honor of his profession is dear, and he will remember that his own character and conduct reflect honor or dishonor upon the profession.

- He will be a good citizen of his country, state, and community, taking an active part in their affairs, especially those dealing with the improvement of aviation facilities and the enhancement of air safety.

- He will conduct all his affairs in a manner that reflects credit on himself and his profession.

- He will remember that to his neighbors, friends, and acquaintances he represents both the profession and ALPA, and that his actions represent to them the conduct and character of all members of the profession and ALPA.

- He will realize that nothing more certainly fosters prejudices against and deprives the profession of its high public esteem and confidence than do breaches in the use of alcohol.

- He will not publish articles, give interviews, or permit his name to be used in any manner likely to bring discredit to another pilot, the airline industry, the profession, or ALPA.

- He will continue to keep abreast of aviation developments so that his skill and judgment, which heavily depend on such knowledge, may be of the highest order.

Having Endeavored to his utmost to faithfully fulfill the obligations of the ALPA Code of Ethics and Canons for the Guidance of Air Line Pilots, a pilot may consider himself worthy to be called...an AIRLINE PILOT."

If you're serious about a career as an airline pilot, the above words bear rereading several times. It's not so much a code of ethics as it is a detailed job description of the airline pilot career. Making life and death decisions are literally what pilots do every minute they're in the cockpit, especially the captain. We seldom think about the possibility of failure, but choose to focus on safe conclusions to the task at hand. To be sure — most pilots are Type A personalities. Trust me, you wouldn't want it any other way.

CHAPTER 6
The Long Journey To The Left Seat

People become airline pilots for a variety of reasons. Some, like myself, can't imagine doing anything else. Other end up in the job after a military flying career, while others just like being abused. Now there's a word that a lot of pilots use to describe their airline careers.

In defense of the industry, "Abused" is a harsh word, but that's the nature of the job.

From the moment you accept employment as an airline pilot your life and the lives of your family members will go through many changes — most can't be imagined by the non-airline families.

For starters, your spouse and children, girlfriend or boyfriend, will see you almost 50 percent less than a normal family. That's the easy part. Here's a list of how a pilot's family differs during the years before they become a captain:

- If you choose to commute you will probably see your family less than your nextdoor neighbor.

- You will miss a majority of birthdays, anniversaries, parties, holidays, christenings, funerals (unless it's a family member), school plays, you son's first touchdown, and your daughter's first dance recital, to name a few.

- You will learn too love airline food, Taco Bell, McDonalds and Burger King and likely gain weight because of you new diet.

- You will now become a constant patron at the dry cleaners. You will want extra starch in your uniform shirts so you can wear them for at least two days.

- Your first few paychecks will have a deduction for the cost of your uniform, hat, tie and shirts. [Cost about $1,000]

- In your first year on probation you might end up working a second job to keep food on the table. Probation pay is getting better, but for many it's still less than $30.00 per hour of flight time, or a salary of less than $25,000 per year. After taxes, medical insurance, etc., some airline pilots are left with less than $1,200 dollars a month of spendable income. This is a tough salary to live on if you have a family.

- Many pilot spouses have to return to the work place during probation year.

- After the first year the wages go up considerably. But remember, that first year contains a large bear trap that's always looming on the horizon:

Even though the Air Line Pilots Association may now represent you, they cannot help a probationary pilot who runs afoul of management during their probation period. Sadly, you cannot take advantage of grievance procedures until you survive the probationary period.

Probation is an unavoidable fact of life when starting your career. In that first year no one cares if you have 10,000 hours of military or commercial flying, a lunar landing to your credit or only a few thousand hours of corporate or private flight time, probation pilots are all treated the same.

Probation is a trial period. It affords both pilot and employer an opportunity to spend time together, testing their compatibility. Because the aforementioned grievance procedures don't apply during this first year, the airline can terminate the relationship without worry about contractual agreements.

Note: Pilot contracts (agreements) often mandate lengthy grievance procedures that must be carried out before a pilot can be terminated. This will only protect you after the probabtion period.

Airline management is not looking for a way to fire a pilot. Airlines hire pilots because they need them. The odds favor your survival of the probation year — but it's still unnerving knowing you're being closely watched.

From year two until the time a pilot checks out as captain it can be a roller coaster ride. Some pilots elect to stay on the same piece of equipment they were initially hired on until they can upgrade. This is the easy path and requires less strain on both the pilot and their family. Others chase the bucks and or the airplanes. Larger equipment typically pays more per hour, but each time you upgrade it requires a lengthy and technically difficult education. This can put your job or ability to upgrade in the future at jeopardy if you fail the final check ride.

Learning to fly a new airplane at this level of aviation is not as easy as you might think. So some pilots negate that risk by staying where they are.

- From what I've observed, most pilots tend to chase the bucks and planes around. I guess we like to suffer.

- Flying new equipment can often mean a change of base. This can result in a family move, or if flying solo, a need to find another pad closer to work. In my career I have lived in Seattle; San Diego; Las Vegas; Houston; Detroit; Niceville, Florida; Ft. Walton Beach, Florida; and Daytona Beach, Florida. All those moves were voluntary and about half were paid for by the airline. Like many other pilots I chased the bucks and larger equipment.

- Each year as a first officer requires annual recurrent training and satisfactorily passing an oral or written exam on the airplane and company flight operations procedures, plus an intense check ride in the simulator. A career 30 plus year airline pilot will likely have around 50 such check rides. I have had 57 as best as I can compute. Trust me, these testing periods are not a walk in the park. I enjoyed the challenge of having my skills tested, but was always relieved when the check ride was over.

- During a normally slow rise through the seniority list, your envy of the captain job usually reaches a point of frustration. Then suddenly your airline increases its fleet size or a large number of senior captains retire. Now it's your turn.

- After initially checking out as captain, my first line flight without a check airman watching from the right seat was an eye opener. Now, I was the guy in charge, the big Kahuna, the man with a plan.

In my excitement to finally be a major airline captain I probably made more mistakes on that first flight than I did the remainder of my career. Nothing major, just silly things like climbing into the right seat out of habit. When the F/O entered the cockpit he said with a straight face, "Hey Cap — wrong seat." Thankfully that's all I remember about my first I'm-finally-a-captain-day.

CHAPTER 7

Low Pay, Long Days And Endless Scrutiny

Numerous chapters in this book reflect on the atrociously low pay pilots must endure early in their career. I've had a great life because of airline flying, but until writing this book I hadn't realized how much of the early years had been purposefully forgotten.

I have spent many nights sleeping in crew rooms after a flight, usually by choice, lack of money, or because of late arrivals and missed commuter flights home. Honestly I feel more at home sleeping in a dilapidated recliner than I do in my own bed, and I love vending machine food.

Airline pilot duty days are longer than eight-hour workdays, a lot longer. Factor in commuting to work the same day as your trip departs and you might find your 'real' duty day can extend to over 20 hours. This is especially true if you're a long-haul pilot. I know this sounds like a model for a fatigue disaster, but you sort of get used to it.

Some pilots like short flights while others love long haul overseas flying. Having done both I have equal likes and dislikes about each.

Long-haul pilots usually make more money and have more days off. The downside is when they go to work it can be for weeks at a time, and they are rarely in contact with their family. Another problem is when you're halfway around the world your circadian rhythm readjusts to the time zone changes after a few days, that's mostly good. Unfortunately when you return home, you're on an opposite sleep schedule from the rest of your family. Again, it takes you several days to readjust your sleep/awake periods. About the time you start feeling normal you go on another trip and the off-schedule sleep patterns starts again. This is bad for anyone's physiology, and a real problem for some pilots. For men it can cause prostatitis; for women, and on rare occasion men, it can also cause fibromyalgia. It seemed like I was always sick when I flew the Boeing 747. I loved being on the plane, but the trips could be physically brutal, especially the west to east all-night crossings of the Pacific.

I spent the last few years of my career flying mostly domestic trips with an occasional trip to Honolulu, Puerto Rico or Mexico. I think as I got older and my waist size expanded the shorter trips were a pleasant relief to my back and bum.

Most 30-year pilots will average between 15 and 20 million miles of flight during their lifetime. That's over 800 times around the earth for 20 million miles of flying. That's a lot of real estate covered.

I live in an airport community (hanger attached to my house and a taxiway behind it) and many of my neighbors and friends are, like me, retired airline captains. Listening to countless stories about their airline days always leaves me feeling good about the glory days I spent flying the line. Despite long hours, lousy meals, constant fatigue, backaches that never seemed to go away and initial slave wages — I'd do it all again.

Four tarnished to perfection captain stripes on a uniform coat that has somehow shrunk over time is the only reminder many of us have of a life above normalcy. It is a long climb to the top. But the view is indescribable.

CHAPTER 8

The 45 Day Ph.D. — A Million Things Not Remembered

As explained in an earlier chapter; the only way a pilot usually gets a pay raise is to upgrade or move to a larger airplane once they've reached the max ceiling for pay, usually 10-12 years.

Upgrading is always a challenging time, the rewards are financial, but often not life pleasing. Let me explain.

Historically most, but not all pilots jump at any opportunity to move up in pay — who doesn't? But usually what happens is they get a pay raise but go to the bottom of the aircraft specific seniority list. That's the good/bad part. In their new position, it often takes years before they can hold a desirable schedule.

Learning how to fly a new jet airliner has been likened to getting a Ph.D. in 45 days. No you don't need to write a dissertation, but you almost need to recite one from memory during an oral that can last for hours, followed by a simulator check ride that usually leaves permanent sweat stains on your clothing.

A typical aircraft-operating manual is as thick as a Manhattan telephone directory. Two to six-week-long ground school classes step you through the aircraft systems, but the trend is for pilots to use electronic media to essentially teach themselves a great deal about the new aircraft before they show up for class. Remember, a jet, is a jet, is a jet. Fortunately, they mostly fly alike, but all jets are not built alike.

CHAPTER 9
Check Rides — Almost Real

If a jet is a jet, what's the big deal about checking out in one? For the Captain to receive a type rating in the large transport jet he or she must demonstrate a working knowledge of each and every control in the cockpit — from memory! It requires hours and hours of study to have a basic mastery of a modern jet. Fortunately, most airlines have developed a building block philosophy regarding airplane instruction.

It used to be that 'ground school' for airline pilots was like drinking water out of a fire hose, but modern teaching tools and techniques have now made the fire hose drink at least manageable.

Most airlines have found that in order to obtain mastery of complex material it's necessary to feed it to the pilot student in small easily digested portions.

Many airplane manuals have been rewritten to reflect this concept. As an example — in older times you'd read a paragraph about flight controls that contained a dozen important need to know facts. It was difficult to remember, or for that matter, comprehend. New thinking breaks down the big-paragraph-many-facts method of teaching into single statements that you can dwell on before moving to the next sentence.

Since computer automation is playing a larger role in flying aircraft, pilots are slowly becoming computer managers. This is both good and bad. Automated flight is so common place that some pilots complain they're loosing their real, mechanical flying skills. A modern autopilot coupled with a flight management guidance computer (FMGC) is a marvelous bit of technology. Type in your flight plan, get airborne, and engage the autopilot; a modern-day airliner can take you all the way to touchdown at the destination. But what happens when the "Magic" fails?

It's easy to have your attention diverted while dealing with "Magic" problems. More than one crash has resulted from pilots failing to fly the airplane while dealing with mechanical or electrical issues.

Programming has become an additional attention grab for pilots. Airline cockpits used to have a third set of eyes we called flight engineers or second officers to work problems while the pilots flew the plane. Modernization and budgetary concerns eventually caused airplane manufacturers to build two pilot cockpits almost exclusively. Most of the time this works quite well, but occasionally it doesn't. Imagine driving down the road at 500 mph, talking to a passenger while reading a technical manual, texting and talking on the cellphone without hitting obstacles in front of you all at the same time. Now add to that that your car's engine just caught fire and smoke is filling up the inside. Hopefully you get my gist.

Flying can be a multitasking nightmare with only two people to deal with emergencies. Training for these types of multi-failure scenarios is not usually done since its been found that teaching the correct response to singular emergencies has proven more effective than having student cockpit crews try and put out several 'fires' at once. Remember "Miracle on the Hudson," where an airliner departed LaGuardia and moments later made a water landing in the Hudson? Fortunately the crew of that plane prioritized and **flew the plane** despite the automation that was slowly degrading because of the dual engine failures.

After weeks of ground school and hundreds of study hours you are given an in depth test you must pass before proceeding to full motion simulator training.

Note: Some airlines now require pilots to show up the first day of training and take a "must pass" airplane systems test.

Usually weeks or months before your training begins, airplane manuals and electronic media like tutorial DVDs and CDs are issued for you to look at on your own prior to the first day of actual ground school. This shifts some of the cost of training to the pilot by reducing costly time the pilot has to spend at the airlines training facility.

After a pilot completes ground school Captains must pass an FAA oral and F/O's must pass an 'in-house' administered oral. The captain will be acquiring a type rating. This is a printed endorsement on his ATP certificate attesting to his qualifications as a captain on that specific type of airplane. The F/O is not normally 'typed,' which is why they don't require the FAA oral.

What is an FAA oral like for a captain? It can be a horrible experience or a walk in the park. It depends mostly on your preparation or lack thereof. Some inspectors are easier on you than others, but all know that passing an unqualified person can lead to unfavorable consequences. The longest captain oral I ever had was when I checked out as a Boeing 727 captain. It was almost seven hours! I had prepared well, too well actually. The FAA examiner told me after it was over that he was new on the 727 and I had answered a lot of questions about the airplane that he was unsure of. That's why he kept the questions coming. I was petrified, while he was getting an education at my expense.

After the oral, the infamous "Type rating," ride follows, sometimes the same day.

A modern day airplane simulator is a marvelous creation. Prior to realistic simulation, all of a pilot's airplane training was accomplished in the actual plane. Usually you'd pick up the airplane after it had finished with the day's passenger flying — that normally meant flying at night. Those glorious days were a hoot as you bored holes through the sky shooting approach after approach with simulated fires, engine failures, etc. Talk about realism! This, as you can imagine, was risky and costly — sometimes even deadly.

Simulators now play a vital role in assuring you, the paying passenger, that your crew has been exposed to inflight emergencies that could never be simulated safely in a real airplane without jeopardizing crew or aircraft. Modern multi-axis simulators are so lifelike in performance and interior looks that after an hour in one you have a hard time thinking of it as a box inside a building.

The instructor sits behind the pilots at a panel that controls almost every indication available to the pilots. A pre-sim briefing (usually one or two hours) covers all details of that training period. Once in the "Box," the period usually starts with a cockpit preflight, reading of checklist, followed by taxi out, takeoff and climb-out. That's when the manure hits the fan. One failure after another is introduced and the crew has to work the problem as if they were actually flying a scheduled trip. Besides working the problem, you're still expected to make calls to ATC, inform the passengers and F/A's on what's happening, and reprogram computers, all the while controlling the airplane, and trying to resolve the problem by (at times) lengthy and complex checklist. These periods are anything but a joyride. The calls I refer to are made through the actual audio control panel — like in the real airplane. Only difference is the instructor plays the roll of ATC and others you might need to speak with.

I sat at the instructor panel for a number of years and learned as much from watching others as they hopefully learned from me. At times the tenor of these simulations is so great, that everyone forgets it's all just pretend.

The type rating check ride is the finale to your training as a captain on a new jet. You've been exposed to major emergencies involving virtually every system on the airplane. The oral is behind you, but that's only half of the check ride. If the type ride is immediately following the oral you might get a short break to compose yourself — perhaps pray. Just kidding, sort of.

Once you enter the box you are to treat everything as though you were on a real flight in a real airplane. You cannot ask for any help other than what you might access in the real world.

The visuals out the cockpit window's are computer generated, but very realistic. For the next two to three hours you will fight engine, depressurization, hydraulic and navigation failures, bad weather, passenger medicals, and my favorite — airplane evacuation. That usually marks the end of the check ride.

Most examiners quietly slip out the simulator's back door when the type ride is over. This leaves you apprehensive. Did you pass or fail? If you made it through the evacuation without the ride being stopped, you're probably golden. If a check ride is going poorly (it happens) the examiner can call it quits at any time. Just so I'm clear on this point. Failing a type rating check ride at this level is embarrassing for the pilot, and for the airline an excess expense. Some airlines still have a two-failures-and-you're fired policy — this is serious business.

Pilots test and take reoccurring check rides as often as twice a year. To make sure we stay on our toes between check rides, we're line checked periodically without notice during actual fights. A check airman, captain from the training department, watches you from a cockpit jump seat as you fly a leg of your trip. All these check rides may seem daunting, and I guess it probably is, but it's all part of the job.

CHAPTER 10
Typical Day Gone Bad

The likelihood of a typical passenger's involvement in an airliner accident is similar to the odds of being struck by lightning. Flying in a modern jetliner captained by an experienced pilot is statistically the safest mode of transportation in the world — even more so than walking.

Like walking, flying sometimes becomes hazardous, even deadly. Despite the billions of dollars spent on aircraft design and training of crews to fly them, occasionally accidents do happen. Some are survivable while others are not.

In over 34 years of airline flying I have never run across a pilot who thought they were invincible —

quite the contrary. What separates them from most people is their overwhelming desire to live. Assuring they have skills to handle most emergencies is a continuous training effort on their part and that of the airline employing them. Can they train for every eventuality? Regrettably, that's a physical impossibility. Pilots train for emergencies that historically have caused most accidents. Unfortunately, that list keeps expanding. Does that mean you, as a passenger need to be concerned? Absolutely not!

Most fatal accidents start with a small problem that rapidly degrades to a major problem. If managed properly in the early stages it can usually be survived. Sometimes, however, dealing with the issue is physically impossible and becomes a good day gone bad.

The following NTSB transcriptions are of radio transmissions of two separate airliners involved in major air disasters. The first transcription is of a survivable crash. The second example ended with 100 percent fatalities.

Some will ask why show these unsettling words? To be candid, it shows good captains and first officers doing what they do best — working the problem and looking for a solution until the very end —

survivable or not. That's what the captain's job responsibility is all about.

* * *

On January 15, 2009 US Airways Flight 1549 an Airbus A320-214 with US registration number N106US departed from LaGuardia Airport and shortly after takeoff suffered bird strikes in both engines. Thrust was lost in both engines and the crew was able to ditch the plane in the Hudson River. All 107 aboard survived.

KEY:
AWE1549 – US AIRWAYS 1549

LGA - La Guardia Tower

ATCl L116 - New York TRACON

LaGuardia Departure TEB - Teterboro ATC

TIMELINE BEGINS:

3:24:54: [Flight 1549 cleared for takeoff]

3:24:58 (LGA): Cactus 1549.

3:25:51 (AWE1549): Cactus 1549 - 700 climbing 5,000.

3:26:00 (L116): Cactus 1549, New York departure radar contact. Climb and maintain 15,000.

3:26:04 (AWE1549): Maintain 15,000, 1549.

3:27:32 (L116): Cactus 1549, turn left heading 270.

3:27:36 (AWE1549): Ah, this is, uh, Cactus 1539. Hit birds, we lost thrust in both engines. We're turning back toward LaGuardia.

3:27:42 (L116): Okay, yea, you need to return to LaGuardia. Turn left heading of, uh, 220.

3:27:46 (AWE1549): 220.

3:27:49 (L116): Tower, stop your departures. We got an emergency landing.

3:27:53 (LGA): Who is it?

3:27:54 (L116): It's 1529. He, ah, bird strike. He lost all engines. He lost the thrust in the engines. He is returning immediately.

3:27:59 (LGA): Cactus 1529, which engines?

3:28:01 (L116): He lost thrust in both engines, he said.

3:28:03 (LGA): Got it.

3:28:05 (L116): Cactus 1529, if we can get it, do you want to try to land Runway 13?

3:28:11 (AWE1549): We're unable. We may end up in the Hudson.

3:28:31 (L116): Alright, Cactus 1549, it's going to be left traffic to Runway 31.

3:28:34 (AWE154): Unable.

3:28:36 (L116): Okay, what do you need to land?

3:28:46 (L116): Cactus 1549, Runway 49-- Runway 4 is available if you want to make left traffic to Runway 4.

3:28:50 (AWE1549): I'm not sure if we can make any runway. Oh, what's over to our right? Anything in New Jersey? Maybe Teterboro?

3:28:55 (L116): Okay, yea, off to your right side is Teterboro Airport.

3:29:02 (L116): Do you want to try and go to Teterboro?

3:29:03 (AWE1549): Yes.

3:29:05 (L116): Teterboro, Empire-- actually LaGuardia Departure got an emergency inbound.

3:29:10 (TEB): Okay, go ahead.

3:29:11 (L116): Cactus 1529 over the George Washington Bridge, wants to go to the airport right now.

3:29:14 (TEB): He wants to go to our airport, check. Does he need any assistance.

3:29:17 (L116): Ah, yes. He, ah, he was a bird strike. Can I get him in for Runway 1?

3:29:19 (TEB): Runway 1, that's good.

3:29:21 (L116): Cactus 1529, turn right 280. You can land Runway 1 at Teterboro.

3:29:25 (AWE1549): We can't do it.

3:29:26 (L116): Okay, which runway would you like at Teterboro?

3:29:28 (AWE1549): We're gonna be in the Hudson.

3:29:33 (L116): I'm sorry, say again, Cactus.

3:29:51 (L116): Cactus, ah, Cactus 1549, radar contact is lost. You also got Newark Airport off your two o'clock and about 7 miles.

3:30:14 (L116): Cactus 1529, uh, you still on?

3:30:22 (L116): Cactus 1529, if you can, ah, you got, ah, Runway 29 available at Newark off your two o'clock and 7 miles.

3:30:30: [Splashdown. Radar and tower notify Coast Guard, which responds, "We launched the fleet."]

3:31:30 (unknown): Was that Cactus up by the Tappan Zee?

3:31:32 (L116): Uh, yeah, it was Cactus. He was just north of the, uh, George Washington Bridge when they had the bird strike.

3:33:38 (L116): Alright, alright. Departure, we're stopped on departure Runway 4 - 360s runway.

3:33:44 (L116): Okay.

3:33:45 (L116): You know about the Cactus?

3:33:46 (L116): Right.

3:33:47 (L116): I guess it was a double bird strike and he lost all thrust, so...

3:33:52 (L116): (Unintelligible) What do you want to do as far as departures?

3:33:55 (L116): Okay, I'll figure it out.

TIMELINE ENDS
– ELAPSED TIME FROM TAKEOFF UNTIL DITCHING:

5 MINUTES 36 SECONDS

* * *

In the mid afternoon hours of January 31, 2000, Alaska Airlines, Flight 261; a McDonnell Douglas MD-83 with US registration number N963AS crashed into the Pacific Ocean south of Point Mugu, California in 650 feet of water. It was en route from Puerto Vallarta to San Francisco. The official NTSB transcription of the radio transmissions from the plane that follow indicate the pilots were struggling with a jammed stabilizer for the last 11 minutes of the flight before nose-diving to a watery grave in the Pacific. During those last minutes they were preparing for an emergency landing at Los Angeles International Airport. Control of the aircraft was eventually lost, and the MD-83 was seen in a nose down attitude, spinning and tumbling in a continuous roll. It was inverted before impacting the ocean. All 88 souls aboard were killed. The probable cause for the accident was loss of airplane pitch control resulting from in-flight failure of the horizontal stabilizer trim system jackscrew assembly's acme nut threads due to insufficient lubrication of the jackscrew assembly.

KEY:
RT = Radio Transmission
CAPT = Captain
F/O = First Officer
FA = Flight Attendant
ME = Mechanic on the ground at LAX
LA14 = LA ARTCC Sector 14
LA25 = LA ARTCC Sector 25
LA30 = LA ARTCC Sector 30

TIMELINE BEGINS:

16:08:03   RT CAPT: Yea we tried everything together.

16:08:08   RT CAPT: We've run just about everything if you've got any hidden circuit breakers we'd love to know about 'em.

16:08:35   RT CAPT: It appears to be jammed, the whole thing, it spikes out when we use the primary, we get AC load that tells me the motor's tryin' to run but the brake won't move it, when we use the alternate, nothing happens.

16:08:50	ME	You say you get a spike on the meter up there in the cockpit when you uh try to move it with the primary right?
16:08:59	CAPT:	I'm gonna click it off you got it?
16:09:00	F/O:	Ok.
16:09:01	RT CAPT:	When we do the primary trim, but there's no appreciable uh change in the uh electrical uh when we do the alternate.
16:09:13	CAPT:	Let's do that.
16:09:14		[Sound of click]
16:09:14	CAPT:	This'll click it off.
16:09:16		[Sound of autopilot disengaging] [Sound similar to horizontal stabilizer in motion tone]
16:09:16	CAPT:	You got it?
16:09:26	CAPT:	It got worse.
16:09:31	CAPT:	You're stalled.
16:09:32		[Sound of air frame vibration]
16:09:33	CAPT:	No no you gotta release it ya gotta release it.
16:09:34		[Sound of click]
16:09:52	CAPT:	Help me back help me back.

16:09:54	F/O:	Ok.
16:09:55	RT CAPT:	Center Alaska two sixty one we are uh in a dive here.
16:10:01	RT CAPT:	... and I've lost control, vertical pitch.
16:10:01		[Sound of overspeed warning] (continues for 33 seconds)
16:10:05	LA30:	Alaska two sixty one uh say again sir.
16:10:06	RT CAPT:	Yea we're out of twenty six thousand feet, we are in a vertical dive... not a dive yet... but uh we've lost vertical control of our airplane.
16:10:10	LA30:	Alaska two sixty one roger.
16:10:20	CAPT:	Just help me.
16:10:28	RT CAPT:	We're at twenty three seven, request uh...
16:10:33	RT CAPT:	Yea we got it back under control here.
16:09:34	RT F/O:	No we don't!
16:10:36	LA30:	Alaska two sixty one uh say the altitude you'd like to uh remain at
16:10:45	F/O:	Let's take the speed brakes off.
16:11:03	LA30:	Alaska two sixty one say your condition

16:11:07	RT CAPT:	Two sixty on, we're at twenty four thousand feet, kinda stabilized.
16:10:55	CAPT:	Ok it really wants to pitch down
16:10:58	RT CAPT:	We're slowin' here and uh, we're gonna uh do a little troubleshooting, can you gimme a block between un, twenty and twenty five?
16:11:21	LA30::	Alaska two sixty one maintain block altitude flight level two zero zero through flight level two five zero
16:11:26	RT CAPT:	Alaska two sixty one we'll take that block we'll be monitoring the frequency.
16:11:43	F/O:	Whatever we did is no good, don't do that again.
16:11:44	CAPT:	No it went down it went to full nose down.
16:11:48	F/O:	Um it's a lot worse than it was?
16:11:50	CAPT:	Yea yea we're in much worse shape now.
16:11:59	CAPT:	I think it's at the stop, full stop...and I'm thinking... can it go any worse...but it probably can...but when we slowed down, let's slow it let's get down to two hundred knots and see what happens.

16:12:33	RT CAPT:	We did both the pickle switch and the suitcase handles and it ran away full nose trim down.
16:12:42	RT CAPT:	And now we're in a pinch so we're holding uh we're worse than we were.
16:13:04	RT CAPT:	(Transmission indicated he was reluctant to try troubleshooting the trim system again because the trim might go in the other direction)
16:13:22	RT CAPT:	I went tab down... right, and it should have come back instead it went the other way.
16:13:32	CAPT:	You wanna try it or not?
16:13:35	F/O:	Uhh no. boy I don't know.
16:14:03	LA30:	Alaska two sixty one uh let me know if you need anything.
16:14:07	RT CAPT:	We're still working at it.
16:14:09	LA30:	Roger.
16:14:54	LA30:	Alaska two sixty one contact LA center one two six point five two they're aware of your uh situation.
16:14:59	RT CAPT:	Alaska two sixty one say again the frequency one two zero five two.

16:15:03	LA30:	Uh Alaska two sixty one twenty six fifty two.
16:15:06	RT CAPT:	Thank you.
16:15:19	RT CAPT:	LA Alaska two sixty one uh we're with you at twenty two five we have a jammed stabilizer and ah we're maintaining altitude with difficulty uh but uh we can maintain altitude we think and our intention is to land at Los Angeles.
16:15:35	LA25:	Alaska two sixty one Alaska Center roger uh you're cleared to Los Angeles Airport via present position uh direct Santa Monica direct Los Angeles and uh you want lower now or what do you wanna do sir.
16:15:56	RT CAPT:	Center Alaska two sixty one I need to uh get down about ten change my configuration make sure I can control the jet and I'd like to do that out there over the bay if I may.
16:16:06	LA25:	Ok Alaska two sixty one roger that stand by there.
16:16:10	LA14:	That's fine go ahead green light
16:16:11	LA25:	Hey Alaska two sixty one wants to go into LA.

16:16:14	LA14:	No problem
16:16:15	LA25:	He wants to get down to around ten thousand feet but he wants to do it out there over the bay.
16:16:17	LA14:	Sure.
16:16:19	LA14:	Ok.
16:16:19	LA25:	I'm gonna send him out on like a two eighty heading right now and then uh.
16:16:22	LA14:	Ok put him on a...
16:16:23	LA25:	He's at two two five right now.
16:16:25	LA14:	Uhh Ok that's fine.
16:16:25	LA25:	Altitude altitude.
16:16:27	LA14:	Put him on a two eighty heading take him down to one seven thousand radar contact.
16:16:29	LA25:	He're we go.
16:16:30	LA14:	(unintelligible)
16:16:31	LA25:	Alaska two sixty one uh fly heading of two eight zero and descend and maintain one seven thousand.

16:16:39	RT CAPT:	Two eight zero and one seven seventeen thousand Alaska two sixty one and we generally need a block altitude.
16:16:44	LA25:	Ok uhh just um I'll tell you what uhh do that for now sir and contact LA Center on three five point five they'll have further uhh instructions for you sir.
16:16:56	RT CAPT:	K thirty five five say the altimeter setting.
16:16:59	LA25:	The LA altimeter is three zero one eight.
16:17:01	RT F/O:	Thank you.
16:17:02	LA25:	Thank you. [This was the last radio transmission from the aircraft]
16:17:04	CAPT:	I need everything picked up... everything strapped down. I'm gonna unload the airplane and see if we can ....we can regain control of it that way.
16:17:09	FA:	Ok we had like a big bang back there.
16:17:11	CAPT:	Yea I heard it.
16:17:15	CAPT:	I think the stab trim thing is broken.

16:17:21	CAPT:	Make sure the passengers are strapped in now.
16:17:24	CAPT:	Cause I'm gonna I'm going to release the back pressure and see if I can get it... back.
16:17:54	CAPT:	Gimme slats extend.
16:17:56		{Sound similar to slat/flap handle movement]
16:17:58	CAPT:	I'm test flying now.
16:18:05	CAPT:	Flaps 11 degrees.
16:18:07		[Sound similar to slat/flap handle movement]
16:18:17	CAPT:	It's pretty stable right now...see but we got to get down to a hundred an eighty.
16:18:26	CAPT:	Ok... bring the flaps and slats back up for me.
16:18:37		[Sound similar to slat/flap handle movement]
16:18:47	CAPT:	What I wanna do...is get the nose up... and then let the nose fall through and see if we can stab it when it's unloaded.
16:18:56	F/O:	You mean use this again?

16:19:01	CAPT:	It's on the stop now, its on the stop.
16:19:04	F/O:	Well not according to that it's not.
16:19:07	F/O:	The trim might be, and then it might be uh, if something's popped back there... it might be mechanical damage too.
16:19:14	CAPT:	I think if it's controllable, we oughta just try to land it.
16:19:16	CAPT:	You think so? Ok lets head for LA.
16:19:21		[Sound of three thumps]
16:19:24	F/O:	You feel that?
16:19:26	CAPT:	Yea
16:19:29	CAPT:	Ok gimme sl....
16:19:33		[Sound of two clicks similar to the sound of slat/flap movement]
16:19:37		[Sound of extremely loud nose and the sound of background noise increasing, which continued until the end of the recording] [Sound of loose articles moving around in the cockpit]
16:19:43	F/O:	Mayday
16:19:49	CAPT:	Push and roll, push and roll.

16:19:54	CAPT:	Ok, we are inverted...and now we gotta get it.
16:20:04	CAPT:	Push push push...push the blue side up.
16:20:16	CAPT:	Ok now lets kick rudder... left rudder left rudder.
16:20:18	F/O:	I can't reach it.
16:20:20	CAPT:	Ok right rudder...right rudder.
16:20:38	CAPT:	Gotta get it over again...at least upside down we're flying.
16:20:49		[Sounds similar to engine compressor stalls and engine spool down]
16:20:54	CAPT:	Speedbrakes.
16:20:55	F/O:	Got it.
16:20:56	F/O:	Ah here we go.
16:20:57.		[End of recording]

TIMELINE ENDS

– ELAPSED TIME FROM TRANSCRIPTION START UNTIL CRASH:

12 MINUTES 54 SECONDS

It's never easy to read the final words of those who struggle to the very end for survival. With totally different outcomes these two examples still had a very common thread — both crews acted exceptionally and as would be expected worked the problem continually — all the way to the water. How is this problem solving different from other occupations? Simply put — flying a jet seven to eight times the max allowable freeway speeds, (approximately 550 mph) miles above earth is pressure enough. Problem solving is now a matter of life and death and needs to be solved in seconds — you can't just pull over, hit pause or have a mechanic take a look at the problem later — everything is important and mistakes are most likely deadly. I can't think of any occupation other than astronaut or test pilot that provides such a work environment.

In Memory

CAPTAIN TED THOMPSON
FIRST OFFICER WILLIAM "BILL" TANSKY

The pilots of Alaska Flight 261 were highly experienced. Captain Ted Thompson (53) had accrued 17,750 flight hours and had more than 4,000 hours' experience flying MD-80s; First Officer William "Bill" Tansky (57) had accumulated 8,140 total flight hours, including about 8,060 hours as first officer in the MD-80. Neither pilot had been involved in an accident or incident prior to the crash.

"Oh, I have slipped the surly bonds of earth,
And danced the skies on laughter-silvered wings;
Sunward I've climbed and joined
the tumbling mirth of sun-split clouds -
and done a hundred things You have not dreamed of -
wheeled and soared and swung high in the sunlit silence.
Hovering there I've chased the shouting wind along
and flung my eager craft through footless halls of air.
"Up, up the long delirious burning blue
I've topped the wind-swept heights with easy grace,
where never lark, or even eagle, flew;
and, while with silent, lifting mind I've trod the high un-
trespassed sanctity of space, put out my hand
and touched the face of God."

— By John Gillespie Magee, Jr.

CHAPTER 11
All Pilots
Not Created Equal

The United States Declaration of Independence provided us with an immortal quotation that decreed, "All men are created equal." This lofty statement unfortunately falls short in describing pilot abilities. All of us would like assurances that the person flying our plane has the ability to do so safely. Unfortunately there is not a common licensing standard shared universally by each individual country. Pilot licensing or certification requirements to fly aircraft are established by the Civil Aviation Authority (CAA) in each country. Each country is responsible for establishing a specific set of knowledge and experience requirements for their citizenship. This includes assurances that the pilot applicant has met minimum standards of airmanship established by observing and evaluating a pilot's skills. The certified pilot can then exercise specific privileges in that nation's airspace. Despite attempts to standardize certification requirements between nations, the current differences serve to limit full international validity of the national qualifications. Here's something to consider — U.S. pilots are certified, not licensed. Legally, pilot certificates can be revoked by administrative action, whereas licensing only requires intervention by the judiciary system; sort of like having your driver's license suspended. Is this important? The simple answer is yes, but I need to qualify this statement for clarity.

In the United States the Federal Aviation Administration (FAA) regulates pilot certification. This branch of the Department of Transportation (DOT) establishes pilot certification standards under the authority of Parts 61 and 141 of Title 14 − Code of Federal Regulations, also known as the Federal Aviation Regulations (FARs). These regulations, in my opinion, are the gold standard of pilot certification criteria. They are not universally accepted or applied verbatim as standard by other countries. This is unfortunate, but before you write-off foreign carriers for your future travel let me explain why universal standards are difficult to implement. This requires a single word explanation: politics.

The freedoms United States citizens enjoy are far greater than any other country on the planet. Learning how to fly in the United States is usually an easy proposition — a freedom we take for granted. The vast majority of countries have, because of political controls, made the process of becoming a pilot quite difficult and extremely costly if funded personally. Learning to fly in a foreign country has become so difficult that from 2009 to 2013, almost 24,000 foreign pilots received FAA commercial or airline transport certificates in the United States. Is this significant?

Air travel in Asian, Middle East and European countries is growing faster than their own pilot resources can support. Because of the short supply of pilots to fill cockpits of these foreign carriers a mad rush to hire pilots has resulted in a lot of inexperienced pilots being put in cockpits of large, widebody jets flying long-haul transoceanic flights. In two-pilot cockpits, having an inexperienced first officer is normally not a problem as long as they're flying with an experienced captain. When things go wrong, well, that's another story.

Over time, all pilots gain experience. Most will learn valuable lessons from this acquired knowledge, and a small number will not. We are, as pilots, definitely not created equal.

Just as this book was going to print, news of the fatal crash of Germanwings Flight 9525 (an A-320 Airbus) dominated the headlines. The crash has been attributed to the deliberate action of its first officer to essentially commit suicide or, more correctly, mass murder. Unfortunately this is the third such incident in recent history of pilot assisted mass murder by a non-U.S. carrier. In 1997 a Silk Air flight from Singapore crashed and the NTSB suggested it might be the result of a pilot suicide. In 1999 a Cairo-bound EgyptAir flight crashed in the Atlantic Ocean off Nantucket, Massachussets, killing 217 people. NTSB investigators concluded the crash occurred because of the co-pilots "manipulation of the airplane controls." The report did not specifically refer to suicide. It's easy to think that avoiding these foreign carriers should minimize exposure to this type of incident. It seems like a logical conclusion, but the reality is far more complex.

As a passenger on any form of transportation there is always a small amount of associated risk. If we didn't take this risk, we'd probably never get out of our beds. The reality of life is that occasionally bad things happen to good people through no fault of their own. The above pilot suicides will likely never be fully understood. As unfortunate and terrible as these events were, they are anomalies. Treat them as such and continue to live and fly to your friends, families and business meetings.

Most likely policy or technical changes will evolve as a result of the Germanwings crash. The airline industry has always reacted to catastrophic events by policy or mechanical changes that minimize and hopefully eliminate reoccurrence of problems.

I would need a super computer to create a statistical model that accurately determines the likelihood of your being a victim in a pilot assisted crash; it's probably a one in a billion flight hour possibility. For the curious, here are some measurable statistics generated by **http://planecrashinfo.com** that you might find interesting.

- From 1950 thru 2010 there were 1,015 fatal commercial aircraft accidents. This sounds like a lot, but represents millions and millions of flight hours. In 2013 the DOT reported 30,057 fatal car crashes in America and 5,657,000 non-fatal crashes. That's for

ONE year. Flying is definitely the safer transportation mode.

— Odds of being on a flight with at least one fatality (as measured by 78 major world airlines): one in 3.4 million.

— Odds of being killed on a single airline flight: one in 4.7 million.

To put this in perspective there are approximately 3.9 million people in Los Angeles. In 2014 there were 264 homicides in this city. By interpretation you are almost 300 times more likely to be murdered in Los Angeles, as you are to die in a commercial plane crash. For those of you brave enough to walk on the streets of Los Angeles, flying is statistically a far safer endeavor.

A quick search on the web shows the vast majority of airline crashes are by foreign carriers. Does this mean they are not as safe as U.S. airline's? Again, this is a yes and no question. Experience is everything when it comes to flying a large jet. In America a flying job with a large carrier is difficult to obtain and only those with the most experience and best credentials are normally selected. The pilot pool in the U.S. is unusually large and fed by our vast network of commuter and regional carriers that provide a constant source of well-qualified individuals for legacy carriers to choose from. Add our former military pilots to the equation and you can see why highly experienced pilots dominate major airline cockpits in America. Unfortunately, many foreign carriers have few well-qualified candidates to choose from and often resort to hiring minimally qualified pilots. Are these carriers safe? Yes, but statistically they are more prone to accidents. Over time, pilot resources and experience levels of all carriers could theoretically equalize. But as demand for air travel increases, countries with limited pilot resources will always struggle to find highly experienced aviators. Meanwhile crash statistics will continue to reveal the obvious.

Below is a list of reported airline crashes from 2014 to March 24, 2015. Do a Web search for airline crash history and with little deduction you can see what the trends are.

2014

- February 16 – Nepal Airlines Flight 183, a de Havilland Canada DHC-6, crashes near Khidim about 74 kilometers southwest of Pokhara, Nepal, killing all 18 people on board.

- February 17 – Ethiopian Airlines Flight 702, a Boeing 767, is hijacked by the co-pilot while en route from Addis Ababa, Ethiopia, to Rome, but lands safely at Geneva. All 202 passengers and crew aboard are unharmed.

- March 8 – Malaysia Airlines Flight 370, a Boeing 777 en route from Kuala Lumpur to Beijing with 227 passengers and 12 crew on board, disappears from radar over the Gulf of Thailand. As of August 1, 2015 a search for the missing airliner continues.

- July 17 – Malaysia Airlines Flight 17, a Boeing 777 en route from Amsterdam to Kuala Lumpur, is shot down over eastern Ukraine, killing all 283 passengers and 15 crew on board in the deadliest civilian airliner shootdown incident.

- July 23 – TransAsia Airways Flight 222, an ATR-72 en route from Kaohsiung to Penghu, Taiwan, crashes during go-around, killing 48 of the 58 people on board.

- July 24 – Air Algérie Flight 5017, a chartered Swiftair McDonnell Douglas MD-83 operating for Air Algérie en route from Burkina Faso to Algiers, Algeria, crashes in the northern

Mali desert after disappearing from radar approximately 50 minutes after takeoff, killing all 110 passengers and six crew members on board.

- August 10 – Sepahan Airlines Flight 5915, a HESA Iran-140 (an Antonov An-140 built under license) crashes shortly after takeoff from Mehrabad International Airport, Iran, killing 39 of the 48 people on board.

- December 28 – Indonesia AirAsia Flight 8501, an Airbus A320 en route from Surabaya, Indonesia to Singapore, crashes into waters off Borneo, killing all 155 passengers and seven crew on board.

2015

- February 4 – TransAsia Airways Flight 235, an ATR-72, crashes into the Keelung River in Taiwan. 42 of the 58 passengers and crew on board are killed.

- March 5 - Delta Airlines Flight 1086, a McDonnell Douglas MD-88 skidded off the runway at LaGuardia Airport and crashed into a fence coming inches from Flushing Bay. No one was killed but more than a dozen people were injured.

- March 24 - Germanwings Flight 9525, an Airbus A320, crashes in southern France en route from Barcelona, Spain to Düsseldorf, Germany. All 144 passengers and six crew on board the aircraft died in the crash.

It's unfortunate that pilot licensing standards differ greatly among countries. But remember it's mostly political anyway and has very little to do with pilot or passenger safety. Remember the jet you're flying on is only as safe as its pilots. Assuming you're on an airworthy aircraft, experience in the cockpit is the most contributing factor to a safe flight, always!

In recent years a growing number of weather related airline crashes headlined the news. If you listen to so-called "Experts" on TV news networks they usually lead you down a confusing path that incorrectly vilifies the pilots; sighting their inexperience, lack of training, time in type of airplane, etc., as probable causes — this humanizes the story. Weather conditions, the obvious "probable cause", are usually mentioned as an afterthought.

Well, here's the truth and nothing but the truth about weather. If you're reading this book you've probably flown in an airliner or perhaps you're reading this book before or during your first flight.

Regardless, most passengers hate bumpy rides and some even have a mortal fear of flight into bad weather. I know this may come as a surprise but some Captains think the same way.

Having a healthy respect for the power of Mother Nature is something all pilots have. Unfortunately many aviators only understand the textbook descriptions of flying in weather and that's not enough.

A good example is surfing. No I'm not crazy — this is as valid an observation as one can make about the forces of nature. Here's what I mean.

If you're near a beach popular with surfers, drive out one day and watch them — even if you don't surf, the sun's vitamin E is good for you. Now that you're at the beach, sit in the sand and watch — you're about to learn one of the most important lessons of flying. All pilots are not created equal.

Watching the surfers you can't help but notice some catch more waves and typically ride these waves longer. Airline captains are like surfers reading the waves — the more experience you have reading the weather the better suited you are in dealing with the sky gods. Like surfers, you learn to read and interpret the elements.

At a furloughed point in my airline career I was the senior research pilot for a government contractor tasked with researching severe weather conditions. For the first time in my flying career I was asked to fly into severe weather, not around it. In reflection I think this was the time in my life I really began to truly understand weather and its potential to do damage to an airplane.

Yes, severe weather can literally overstress the airframe of a modern jetliner causing an in-flight break-up. The trick to preventing this is reading the weather, and avoiding those severe conditions. I use the term "reading" to describe the process of evaluating weather.

In my career I've had passengers poke their head in the cockpit and state on more than one occasion, "Just wanted to see if there was any gray hair in the left seat." I'm glad now that I never dyed my hair!

Humor aside, most of you will never see your cockpit crew, much less the color of their hair. Truth be told, we all secretly hope the guy or girl captaining your plane is the most experienced pilot in the world. Even I feel that way as a passenger.

Experience flying in bad weather is just that — experience. Nothing can replace exposure to severe elements like time in the cockpit dodging bad weather.

If a captain has hundreds of hours flying through rain, sleet, snow, ice, and around thunderstorms and tornadoes — they will usually be much better weather pilots than those with limited exposure to bad weather.

Is this to say you're safer with an experienced weather captain when navigating through or around weather? I know I'm going to get criticized for this answer — but yes, if you fly in bad weather with an inexperienced weather crew you are at a slightly greater risk of being a victim of a weather related air disaster. Before the entire population gets up out of their tiny seats in small commuter planes piloted by 20-something's — let me qualify that statement.

The basic criteria for holding an Airline Transport Certificate — ATP (required license to fly as pilot-in-command — PIC of a large airliner over 12,500 pounds gross weight,) is the same for commuter captains flying 50 passenger planes as it is for a Boeing 777 captain flying 450 people. An in-depth FAA written that covers advanced weather knowledge, and a myriad of other aviation subjects, plus an exhausting airplane check ride must be passed before being issued an ATP certificate. So why aren't the weather skills equal if you have an ATP? To answer that lets take a look at the current proposed requirements for obtaining an ATP.

Pilots pursuing an ATP certificate after July 31, 2014, must have logged a minimum of 1,500 hours, complete an ATP certification training program consisting of 30 hours of ground school plus 10 hours of simulator training prior to being eligible to take the ATP written and practical tests. The 10 hours of simulator training will include six hours of training in a level C or D (full-motion) simulator. This differs somewhat from the old criteria with addition of the ground school and simulator training requirement being new.

The new rule establishes a new ATP certificate with restricted privileges for piloting multiengine air-planes only. This restricted ATP can only be used to serve as a first officer at an air carrier. To obtain that certificate an applicant must be at least 21 years old, hold a commercial pilot certificate with instrument rating, complete an ATP certification-training program, and pass the ATP written and knowledge tests. Restricted ATP applicants do get some relief — they are required to have at least 750 hours total time as a military pilot; at least 1,000 hours total time and a bachelor's degree with an aviation major; or at least 1,250 hours total time and an associate's degree with an aviation major; or 1,500 hours total time as pilot.

A flight instructor teaching private pilots to fly light single engine airplanes can easily accumulate 1,500 hours flight time in a year and a half — with very little, if any, exposure to heavy weather. That person could legally be at the controls of your airliner during a winter storm.

The one distinction the FAA ATP criteria fails to address is type of flying or equipment used in acquiring those first 1,500 hours of flight time. You can have heavy transport time from the military or light airplane time like our flight instructor example.

What's all this have to do with weather captains? Everything!

Experience in an occupation is what separates amateurs from pros. Flying jetliners in bad weather, as I've hopefully pointed out is not automatically assigned to the most skilled pilots. Airline crews bid schedules based on seniority and days off — weather flying is not part of the trip selection process since weather forecasting is still somewhat of a black art.

Older pilots, unless they started late in life with a career change, are less likely to make judgment errors when flying into rough weather. That's the simple truth, they simply have more experience in weather analysis and avoidance. As the saying goes, "You can't buy experience."

Again, to ward off panic and to keep you from jumping ship, smaller carriers (regionals) typically have very talented but sometimes under experienced crewmembers manning the cockpit. They are not unsafe weather pilots anymore than a new doctor is an unsafe medical practitioner. To their credit they usually gain experience rapidly due to rigorous "up and down" flight schedules. If they fly in seasonal bad weather country, they shoot far more instrument approaches than the big carriers and get exposed to a lot more weather. This is all good. Additionally regionals are much more conscious about crew pairings than they have been in the past. Matching skill levels (inexperienced pilots flying with more experienced aviators) is an ongoing evolution for carriers and regulation makers.

Current FAA regulations require the Pilot-In-Command (PIC) to make all takeoffs and landings in certain situations if the Second-In-Command (SIC) has fewer than 100 hours of flight time in operations under part 121 (Regulations for Air Carrier Certification) in the type airplane being flown. Additionally, current regulations prohibit operations under part 121 unless either the PIC or the SIC has at least 75 hours of line operations flight time in the type of airplane being flown. What does this mean to you? For starters it resolves some issues regarding experience levels in the cockpit. Know this, all pilot's start out with zero experience especially where weather flying is concerned. New pilots entering airline ranks are watched carefully as they develop and airlines and the FAA are doing all they can to assure that your crew is up to the job.

Unfortunately I've had the unpleasant experience of flying with pilots who seem frightened to death at the prospect of flight into bad weather. Most of that fear is attributable to lack of exposure. In almost 40 years of commercial aviation, I have spent several thousand hours flying into weather. Strangely enough the roughest rides I've ever encountered were in clear air. You can't avoid what you can't see — that's almost true.

All major US airlines have superb dispatch and meteorology departments that can contact a plane in flight, almost at will. These folks are the shepherd's tending the weather. With their assistance, flights can usually be given smoother route alternatives. In addition, pilots frequently give 'ride reports' that assist others in selecting routes and altitudes for a smother ride. These so called "ride reports" are often times the best source of in-flight weather information.

Sometimes, despite our best efforts it's impossible to find smooth air. Pilots often joke among themselves labeling those unavoidable bumpy rides as, "Passenger appreciation flights." Honestly we don't like bumps either — it spills our coffee.

Here's the rub about weather avoidance; sometimes captains have a lapse in judgment or wait too long before asking for help from dispatch or air traffic control — regrettably they can become trapped in a weather system with no smooth way out.

In recent years some weather related accidents could likely have been avoided had the Captain taken action sooner. Unlike unexpected earthquakes — bad weather and rough rides usually provide telltale warning signs. Unfortunately there is no practical way too measure weather flying experience. Written test or check rides can simulate or evaluate your weather knowledge. But nothing about Mother Nature lends itself to definitive interpretation.

Moving at over 500 mph requires thinking way ahead of the aircraft. For some Captains that's easy. For others — well lets just say it's a constant struggle to keep up with the airplane. Fortunately these types are few in number, but they occasionally sneak under the radar and into the left seat — at the major airline level they are usually identified quickly and retrained or terminated. Remember new hires with major airline are usually on probation for a year — during that time those without the right stuff are usually terminated. It's a brutal but qualifying first year.

CHAPTER 12
Conflict In The Cockpit

Do pilots argue in the cockpit? Well, we'd prefer not to but it happens.

Throughout a pilot's career they deal with varying levels of conflict almost on a daily basis. Decisions will be required, and at times they will not be agreeable to everyone. This is the part I don't miss. Most retired captains I spoke with about this opined the same response. We hate conflict, period!

I can't think of an office with more potential for conflict than a modern airline cockpit. Here's why. Lock yourself in a closet sized office with someone who's normally opinionated (as are most pilots) with a Type A personality, and eventually, despite your best efforts to get along or agree, it just doesn't happen. It's human nature. We don't always like or concur with the decisions of others — especially when they might put your life in jeopardy. Fortunately most flight deck crews seem to work hard at avoiding this "showdown" of opinions.

I've said elsewhere that getting along is a necessary component of the job. So this section is directed more toward the aspiring captains among you.

My wife doesn't understand why I usually say little in an argument. The simple truth is she's probably right, but regardless of the debate, no disagreement is worth injuring a happy relationship. In a manner of speaking the interaction between a captain and crew is best served when treated like a happy marriage.

Captains are paid to be calm, methodical problem solvers. While not all achieve such status, we at least try.

Even the worst captains I flew with knew when to stop arguing and dial down the 'I'm-acting-like-a-jerk' meter. I remember being responsible for most of the defusing. No, I'm probably not that cool, but over the years I have gained some smarts about dealing with turbulent personalities. I was fortunately never exposed to that much conflict in the cockpit. But many are.

The naked truth is, some people in this world just have difficulty in getting along. Having a chip on your shoulder is fortunately something that rarely happens in the airline cockpit, but it does happen.

Professionally speaking, most pilots at legacy carriers are not only intelligent, but wise enough to detect personality problems early on. If pilots see that a potential conflict in personalities might jeopardize a safe operating environment one of them will usually say something. Remember all pilots have certificates that can be revoked by the FAA. The possibility of losing this hard-to-get job is a good incentive to behave. If a clash starts while sitting at the gate the quick simple cure is to leave the cockpit and get a replacement crew. This usually means an unpleasant trip for you both to the Chief Pilot's office.

Here's advice for all of you in the work place (especially pilots) from a captain that's been there, done that, and lived to write about it:

There are two sides to every argument so choose the 'livable one' that makes the other person happy, it's a victory for both of you.

- The pro listens, the fool argues — be one (you choose) not the other.

- Practice of good people skills isn't just for captains.

- Confrontations are for hostile individuals, sincere discussions are reserved for concerned professionals — which one are you?

- Better to win a friend than to lose their respect. This doesn't mean agreeing to an unsafe proposal. If you know you're right — convince the other pilot. All pilots are logical thinkers. Sometimes we just need to put the disagreeing party in focus.

- The other pilot can save your life, so why chance pissing them off. A good crew can count on each other when things "get busy".

- The smarter person may win a senseless battle, but the wiser person allowed them to do so.

Conflicts in the cockpit can and do happen. Keeping your character in check is a sign of maturity.

Taking the easy and safe way out of a confrontation takes courage and compassion. Arm yourself with both.

Many years ago I was called from the flight deck to defuse a drunken brawl between several First Class passengers. I put on my coat and hat before leaving the cockpit, hoping the authority figure presence of yours truly would help. The entire First Class section listened as I calmly explained what I was going to do if the drunken individuals did not remain quietly in their seats for the duration of the flight. The threat of being arrested upon landing seemed to get most of their attention. Satisfied things were under control I turned back toward the cockpit but felt a tug on my coat sleeve. I looked down and saw a well-dressed man motioning to speak with me. I leaned down and he whispered, "How come you're so calm?" I simply answered, "That's what I get paid for."

For some reason pilots seem to have a better grasp on dealing with confrontations. Maybe it's because of our 'tiny-locked-in-a-box' occupation. Or perhaps it's because we carry an axe and firearms on most flights. Regardless, a conflict can be dangerous in the cockpit. If you want to be a captain and this is an issue, you might consider keeping your occupation choices on the ground.

CHAPTER 13

Where The Buck Stops And Why

The airline captain is responsible for overall safety of a flight. What exactly does this statement mean?

Nothing makes an airline pilot's blood boil more than reading about or watching a news cast erroneously, not to mention irresponsibly, vilify a pilot after an airline crash. Accidents are just that, accidents.

In May 2003, Capt. Lindsay Fenwick (Northwest) and Michael Huhn, ALPA Senior Staff Engineer wrote the following in *Air Line Pilot* magazine:

"An accident, by definition, is an occurrence that is not expected, foreseen, or intended. By this definition, then, an accident is not a crime. In English Common Law, a crime requires two elements—intending to commit an unlawful act and actually committing the deed. Most airline accidents are just that—unfortunate and unforeseen consequences of human error, not necessarily the pilots'."

Punishment, real or threatened, can have an adverse effect on the conduct and quality of accident investigations. So why then do we always seem to spotlight pilots as being the bad guys? The casual answer is because pilots are at the 'pointy' end of the plane and usually the first to feel the impact of a crash. The correct answer is: the buck does stop with (you guessed it!) the Captain. They are always the individuals with ultimate accountability.

Lets take a look at what happens after a crash as it pertains to ALPA represented pilots.

Surviving flight crews are advised to provide written statements and interviews to the NTSB and other organizations. This information will become public record and eventually obtained by both the employer and regulator (the FAA). Consequences of involvement in an accident can include company discipline (remedial training, unpaid leave, or dismissal) and/or certificate suspension or revocation, fines, etc.

Here's the unfortunate part of most airline accident investigations. It's a given that statements of surviving flight crewmembers are critical to accident investigations. In the United States, when a pilot acts in what they perceive to be in the best interest of safe aircraft operation, the assumption is they will not be criminally prosecuted. Based on this reliance, and in spite of the potential negative effect of full disclosure, ALPA pilots are still encouraged to cooperate fully and aid in accident investigations.

Dead pilots make great defendants for prosecutors seeking wrongful death compensation for plaintiffs. Captains on legacy carriers are very experienced and extremely cautious. Sure they make mistakes, but never (to my knowledge) purposely. That unfortunately does not preclude attorneys from trying to prosecute them.

Many accidents are a result of operational or mechanical issues that are not resolvable in flight. Accident investigation has played a vital role in identifying mechanical, operational, and human factors issues that need to be addressed. This investigative process, and the resulting knowledge gained from catastrophic accidents, enables us to prevent their recurrence. This is what an airline accident investigation should be about! Not placing blame on the crew, but preventing a recurrence.

Nobility aside, all pilots realize the hazards of this occupation. We have no desire to hurt or kill a passenger. We are always the first in the court of public opinion to be singled-out as possible cause for a crash, Is it right, is it fair? Probably not, but it comes with the job. Rather than establishing blame, causal factors should be identified, safety recommendations made, and appropriate corrective actions taken.

News reporters take note: Pilots are the good guys. It's the system of unjustified blame that vilifies us and slows the healing process after an air disaster. Yea, we'll gladly take the fall if it means saving lives in the future, that's why the buck stops with us!

CHAPTER 14

Cockpit Resource Management — A Long Time Coming

In the beginning there were Captains — gods by their own admission. What came from the mouths and actions of these deities was considered gospel. Co-pilots and flight engineers walked sheepishly in the shadows of these four striped, cockpit monarchs. That was then — this is now.

When I started working for my first airline it was not as a pilot but as a mechanic. I was 19 years old and figured I could attend college by day and manage a full-time job at night slinging wrenches on airliners. I guess that's when I got used to long hours and lousy food. It's also where I got introduced to some unique captains. I love analyzing personalities, and in the 60s there were still a lot of huge personalities captaining airliners.

It still remains gospel that captains have final say on the operation of their aircraft, but there's a distinct difference in the way decisions are now made in airline cockpits.

Crew resource management (CRM), known by other names such as cockpit resource management, and flight-deck resource management, is a term coined by NASA psychologist John Lauber. For years he studied communication processes in cockpits. The god in the left seat mentality was observed to be alive and well. But, one thing became obvious to Lauber; better communication between crewmembers could greatly enhance cockpit safety. Goodbye god — hello team player.

Lauber proposed making cockpit culture less authoritarian. Co-pilots and flight engineers would be encouraged to question captains if they observed them making mistakes. This was not easily accepted by some of the older captains — but they would eventually come to accept the concept, albeit grudgingly.

CRM came about after analysis of the 1978 crash of United Airlines Flight 173. (They ran out of gas while troubleshooting a landing gear problem.) After the crash investigation the NTSB made a landmark recommendation to require CRM training for airline crews. In my humble opinion this is one of the best recommendations to ever come from a government agency.

By the early 70s I had already obtained my first airline job as a pilot and got to see first hand the Captain-is-god scenario played out. Looking back, I realize that these old style captains were self-contained in their commands. By that I mean that a great deal of the time they were literally on their own. Communication in the cockpit or to outside resources was never encouraged since most airlines were only concerned about their pilots keeping on schedule and not breaking the equipment. That giant job was dumped squarely on the captain's shoulders. I guess I understand why some of these guys were jerks, they had to be, considering the environment they operated in.

In 1981 United was the first airline to provide CRM training for cockpit crews — they had no desire for a repeat of Flight 173. By the 1990s CRM training for aircrews had become global.

If you want to see a good example of the necessity for CRM take a look at the 1954 classic John Wayne movie "The High and the Mighty." The "Duke" portrays a very vocal washed up co-pilot that ends up saving the day on a trans-Pacific flight. This is probably the first film example of what CRM would later become. Thanks "Duke."

With acceptance of CRM concepts the captain's role in the cockpit has necessarily been redefined.

Yes, the captain is still the final say, but here's why their decisions are now, shall we say, more informed.

One of the first problems CRM education faced was defining how cockpit communication could be more effective. Some more vocal F/O's and engineers thought they now had equal say in the command of the aircraft. Sadly this brought about more than one heated conversation in the cockpit. The intention of CRM is to allow all crewmembers freedom to speak their mind and challenge what they interpret as a bad call by the captain. As you can imagine this didn't always come about as intended. It took a while before everyone got on board and understood that in order for cockpit communication to be effective all parties including the captain must handle this new philosophy delicately. Over time crews have adapted and captains have learned to share the burden of command that ultimately leads to better decision making. They're still the boss, but they're now surrounded by tactful advisors.

I've never met an airline captain who didn't have an entertaining story about a mistake they'd made. Most were silly forgetful things, while other events were more serious, but ended on a high note. Regardless, the cockpit of an airliner is where mistakes should be as few as humanly possible.

CRM concepts have spread out of the cockpit and into many totally different cultures. A close friend of mine is a surgeon and with approval of one of his patients, invited me into the OR to observe a complicated surgical procedure and to see how he worked. This was over 20 years ago. He was terrific to watch, but the assisting surgeon was demanding, abusive and was clearly viewed with contempt by the scrub nurses and everyone else in the OR. After the surgery, "Wow," was all could say to my friend. Talk about different types of surgeons. I asked my friend if hospitals or even medical schools provided training on resource management. He had no idea what I was talking about. Since then many hospitals have initiated resource management for surgeons and OR staff. That call was greatly needed.

Here's my point. Captains have become more effective with adoption of CRM principles. Truthfully it's a relief to know that the airlines are now fully on board with CRM. They remind captains (and other crew members) at every opportunity of all the resources available to aid them in the decision making process. Two or three brains working together in the cockpit are far better than one. Now factor in the entire airline and FAA resources that can assist and you have a powerful problem-solving tool.

I sort of miss being a god, but to a greater extent I appreciate what CRM has done for Captains and ultimately for you the passenger.

For those inclined to think that the captain is the only one flying the plane and making all the occasionally death stopping decisions — think again.

There's still only one captain, but he or she now takes advantage of many different brains.

CHAPTER 15

Management Versus Flight Crews

I have been in airline management, a line pilot, check airman, instructor pilot, and faithful employee for more than one airline over many years. This in itself is not remarkable. Working for these different airlines provided me a broad perspective about differing airline cultures.

Major airlines, and smaller carriers, are complex entities requiring many different specialists to put an airplane in the air. Atop all these specialty departments, mid- and high-level executive managers keep their staff in sync with the rest of the airline. As can be imagined more technical departments like maintenance and flight operations consume the lion's share of management time.

Many years ago I was hired as director of flight operations for a small airline operating DC-9 airliners. During the interview I was told the company's pilots needed babysitting and discipline when not 'keeping with the program.' Despite this proclamation from a senior vice president, I took the job. Sadly I found he was correct. As it turns out pilots are just like other employees. Yes, they might be better educated, earn more money and have the respect of fellow airline employees, but deep down they're just people sharing the same burdens and problems we all share. Sitting in my new office and not the cockpit gave me a new perspective and appreciation for management.

For some airline managers the perception that pilots are underworked, overpaid, spoiled brats still exist. Fortunately that's not an opinion shared by all.

The average major airline captain salary is around $150,000 a year. In 2011 the Associated Press reported the average airline CEO's income was nearly 3.5 million. Quite a pay disparity considering the pilot is actually the one making money and the CEO is the one spending it.

I sincerely believe these high paid executives probably deserve a high income but historically it has been earned from the sweat of others.

As a manager it gave me perspective about a constant mini-battle that seemed to exist between management and flight crews.

Somewhere along the way much of airline management seems to have lost sight of simple facts. Despite their Wharton and Harvard business school educations some fail to grasp the important contribution pilots make to the airline. Pilots are the one making the money. No pilots, no flying, no income for the airline, and then ultimately no airline. To be fair, this situation is changing.

A number of airlines have closed their doors on account of pilot pay issues during contract negotiations. This is regrettable. Now everyone suffers, but especially pilots. Good managers and ground personnel can usually find airline industry jobs, but pilot jobs are not so easily obtained. Add to that, older captains looking for a job will be competing with much younger aviators who can serve the airline longer. Regrettably that always places pilots in a lower bargaining position. Contract negotiating time has historically been a period when the management versus pilot relationship is felt within the airline; this is why last minute negotiation agreements are often arbitrated by an independent third party. Large airlines ceasing operations can have a devastating impact on a nation's economy.

To understand the captain's job or for that matter the F/O or engineer job you need to put the position into the airlines hierarchy, or pecking order. Simply put:

- Pilots are highly trained assets that the airline spends hundreds of thousands of dollars to train and retrain during their career. Officially, you are also an expense item on a financial spreadsheet.

- While it's true, without pilots airplanes don't fly, it's also true that without creative and financially savvy managers running the airline, you will never fly a $400 million jet.

- Bottom line: management and pilots need each other. Pilots probably being the more needy of the two groups. But, that's because we are only employed because some manager type way-back-when decided to start an airline. There will always be somewhat of a contentious relationship between the two factions. It is far better than it used to be — who knows what the future holds. I can't imagine what a pilotless airline contract will be like.

CHAPTER 16

Memorable Moments Of Flight

In the almost 50 years that I've been a pilot I have met and flown countless movie stars, Miss America's and her runner-ups, Rock and Roll Hall of Famers, most NFL, NBA, NHL teams, famous models, politicians, etc. On the flip side of these exciting encounters, I have experienced strikes, low pay, long hours, missed anniversaries, birthdays, and have slept in crew room recliners for more nights than my back cares to relive. The one story that stands out is what happened on and shortly after 9/11.

On the early morning of Sept. 11, 2001 I was in Minneapolis at my airlines' training center getting a check ride in a Boeing 757-200 simulator. It was still dark when I entered the simulator at 6 a.m.

At the halfway point of the ride (8 a.m.) we took a break to stretch our legs. At the reception area to the training center a small crowd had gathered around a portable TV and everyone watched as looped footage played over and over showing an American Airlines Boeing 767 crash into the north tower of the World Trade Center.

No one spoke — all eyes stared at the tiny TV screen as the drama unfolded.

Eighteen minutes later a United Airlines Boeing 767 crashed into the south tower near its 60th floor. I don't know if it's possible, but I swear not one of us breathed during that 18 minutes.

After the second crash the check pilot reminded us that the F/O still had to complete his half of the check ride. None of us were in focus, but somehow we managed to satisfactorily complete the ride.

What happened in the aftermath of the 9/11 tragedies was devastating for every U.S. citizen, but airlines in particular were most affected, especially from an economic standpoint. Many people decided flying would no longer be safe and let those thoughts be known in a big way. Airline ticket sales plummeted to a nearly nonexistent level — more importantly, trust in the airlines to keep passengers safe was perceived as insufficient.

Almost immediately following the attack all aircraft over, or approaching U.S. soil were ordered to land. The ensuing chaos was an expensive nightmare for the airlines, and unimaginable inconvenience for passengers scattered across the landscape. The rental car companies were making a fortune, while some airlines were depleting lines of credit to dangerous levels just to keep the doors open.

For those too young to know or getting too old to remember, the following is a timeline of events surrounding the 9/11 attacks.

The crewmembers lost that day are a not-so-subtle reminder of the ultimate sacrifice American and United Airlines pilots and flight attendants made.

American 11 (Boston to Los Angeles)
Crashed into World Trade Center

John Ogonowski, Dracut, Mass., Captain

Thomas McGuinness, Portsmouth, N.H., First Officer

Barbara Arestegui, flight attendant

Jeffrey Collman, flight attendant

Sara Low, flight attendant

Karen Martin, flight attendant

Kathleen Nicosia, flight attendant

Betty Ong, flight attendant

Jean Roger, flight attendant

Dianne Snyder, flight attendant

Madeline Sweeney, flight attendant

United 175 (Boston to Los Angeles)
Crashed into World Trade Center

Victor Saracini, Lower Makefield Township, PA, Captain

Michael Horrocks , First Officer

Robert J. Fangman, flight attendant

Amy N. Jarret, flight attendant

Amy R. King, flight attendant

Kathryn L. Laborie, flight attendant

Alfred G. Marchand, flight attendant

Michael C. Tarrou , flight attendant

Alicia N. Titus, flight attendant.

American 77
(Washington/Dulles to Los Angeles)
Crashed into the Pentagon

Charles Burlingame, Captain

David Charlebois, First Officer

Michele Heidenberger, flight attendant

Jennifer Lewis, flight attendant

Kenneth Lewis, flight attendant

Renee May, flight attendant

United 93 (Newark to San Francisco)
Crashed in Shanksville, Pennsylvania

Jason Dahl, Colorado, Captain

Leroy Homer, Marlton, N.J., First Officer

Sandy Bradshaw, flight attendant

CeeCee Lyles, flight attendant

Lorraine Bay, flight attendant

Wanda Green, flight attendant

Deborah Welsh, flight attendant

9/11 Timeline of Events

8:00 a.m.: American Airlines Flight 11, carrying 81 passengers and 11 crewmembers, begins its takeoff from Logan Airport in Boston en route to Los Angeles.

8:13 a.m.: An air controller instructs American 11 to climb to 35,000 feet, and the aircraft fails to respond. The controller tries using the emergency frequency to contact the aircraft but there is no response.

8:14 a.m.: United Airlines Flight 175, carrying 56 passengers and nine crewmembers on board, takes off from Logan Airport in Boston.

8:20 a.m.: American Flight 77 takes off from Dulles International Airport.

8:21 a.m.: American Flight 11 turns off its transponder.

8:24 a.m.: This transmission from a hijacker comes from American 11: *"We have some planes. Just stay quiet, and you'll be okay. We are returning to the airport"* ("We have some planes," was unintelligible.) Seconds later another statement follows: *"Nobody move. Everything will be okay. If you try to make any moves, you'll endanger yourself and the airplane. Just stay quiet."* (Most likely the terrorist thought he was talking to the passengers, not transmitting outside the aircraft.)

8:28 a.m.: The FAA's Boston Center calls FAA Command Center in Herndon, Virginia, and reports that American 11 has been hijacked.

8:34 a.m.: Boston Center receives another transmission from American 11: *"Nobody move please. We are going back to the airport. Don't try to make any stupid moves."* (Terrorist presumably still thinks he is giving PA, not transmitting outside the aircraft.)

8:37 a.m.: Boston Center informs NORAD of American 11's hijacking. It is the first notice the military receives of the unfolding events.

FAA: *"Hi. Boston Center TMU, we have a problem here. We have a hijacked aircraft headed towards New York, and we need you guys, we need someone to scramble some F-16s or something up there, help us out."*

NORAD: "Is this real-world or exercise?"

FAA: "No, this is not an exercise, not a test."

8:41 a.m.: United 175 enters New York airspace.

8:42 a.m.: United Flight 93 takes off from the airport in Newark, New Jersey.

8:46 a.m.: In response to American 11's hijacking, Otis Air Force Base receives an order to scramble F-15s. Military officials ask for a destination for the fighter planes. None is known. At the same time of the order, American 11 crashes into the World Trade Center North Tower.

8:47 a.m.: United 175 changes its transponder code twice. The changes go unnoticed because the same controller assigned to it is looking for American 11.

8:48 a.m.: The FAA's New York Center, unaware of American 11 crashing, talks to the FAA Command Center in a teleconference concerning that flight. *"This is New York Center. We're watching the airplane... They've told us that they believe one of their stewardesses was stabbed and that there are people in the cockpit that have control of the aircraft..."*

8:50 a.m.: The military receives word that a plane has hit the World Trade Center. At the same time, American 77 ceases communication with air traffic controllers.

8:51 a.m.: The controller notices a change in the transponder code from United 175. The plane does not respond to repeat requests to change it back.

8:53 a.m.: Military fighter jets summoned for American 11 are airborne, but still lack a target and information about the threat. At the same time, the air traffic controller tells a peer there is a second hijacking and United 175 is unaccounted for. The controller begins diverting planes from the path of United 175.

8:54 a.m.: FAA Indianapolis air controllers notice American 77 deviating from its flight plan. The flight does not respond to contacts, and controllers are unaware of the hijacking and crash in New York.

8:56 a.m.: American 77 turns off its transponder.

9:00 a.m.: FAA Indianapolis notifies agencies that American 77 is missing, possibly crashed and seeks military help for a search and rescue.

9:01 a.m.: FAA New York Center notifies its command of its dilemma. *"We have several situations going on here. It's escalating big, big time. We need to get the military involved with us."*

9:02 a.m.: FAA New York Center asks New York's terminal approach for help in finding United 175.

Terminal: *"I got somebody who keeps coasting, but it looks like he's going into one of the small airports down there. Got him just out of 9,500 -- 9,000 now."*

Boston Center: *"Do you know who he is?"*

Terminal: *"...We don't know who he is. We're just picking him up now."*

Boston Center: *"Heads up man, it looks like another one coming in."*

9:03 a.m.: United Airlines Flight 175 crashes into the South Tower of the World Trade Center. FAA Boston Center deciphers the message from American 11, and realizes hijackers control more than one plane.

9:08 a.m. til 9:13 a.m.: Military fighters are in a holding pattern over Long Island's coast. NORAD decides to ask the FAA to enter New York airspace as a defensive move. *"... If this stuff is gonna keep on going, we need to take those fighters, put 'em over Manhattan. That's the best play right now."*

9:10 a.m.: American 77 enters FAA Washington Center space but goes undetected for 36 minutes as the FAA checks westerly points for the craft.

9:09 a.m.: Fighter jets in Langley, Virginia, move to battle stations to backup New York fighters who may get low on fuel.

9:17 a.m.: The Federal Aviation Administration shuts down all New York City area airports.

9:20 a.m.: FAA Indianapolis learns about the other hijackings and becomes suspicious about American 77.

9:21 a.m.: The Port Authority of New York and New Jersey orders all bridges and tunnels in the New York area closed, and FAA Command Center tells Dulles terminal to look for targets.

9:25 a.m.: The military establishes a combat air patrol over Manhattan, and the FAA issues a nationwide ground stop of all aircraft.

9:28 a.m.: The FAA receives its last normal communication from United 93.

9:29 a.m.: FAA Cleveland Control Center hears screams and struggles from an unknown source and someone yelling, "Get out of here! Get out of here!" The control center notices United 93 has dropped 700 feet.

9:30 a.m.: President Bush, speaking in Sarasota, Florida, says the country has suffered an "apparent terrorist attack." Cleveland Center polls other flights to determine if they heard the screaming at 9:29. Several report they did.

9:32 a.m.: Dulles terminal spots a suspicious aircraft and notifies the Secret Service. An unarmed National Guard cargo plane begins following American 77. Cleveland Center receives another transmission on the frequency from where there was screaming. "Keep remaining sitting. We have a bomb on board." (The transmitter selector panel, located on the center pedestal of the Boeing 757/767 series of jets, is a maze of pushbuttons, switches and glowing indicator lights — this and other inadvertent transmissions by the terrorist were likely caused by incorrect use of the selector panel.)

9:34 a.m.: FAA notified that United 93 might have a bomb on board. Until 10:08, Cleveland Center provides FAA updates on United 93's course.

9:36 a.m.: NORAD learns of a suspicious aircraft a few miles from the White House and orders the Langley fighter jets back to Washington. Cleveland Center asks whether anyone has requested military interception of United 93.

9:38 a.m.: American 77, with 58 passengers, four flight attendants and two pilots, crashes into the Pentagon, the National Guard pilot reports the crash to Washington's terminal facility. The Langley fighter jets are 150 miles away.

9:39 a.m.: A radio transmission from United 93 crosses. It is the voice of hijacker Ziad Jarrah: *"Uh, is the captain. Would like you all to remain seated. There is a bomb on board and are going back to the airport."*

9:41 a.m.: FAA Cleveland Center loses United 93's transponder signal, but uses visual sightings from other planes to track its turn east, then south.

9:42 a.m.: FAA Command Center learns from television reports that a plane has struck the Pentagon. The FAA orders all airborne craft to land at the nearest airport.

9:45 a.m.: The White House is evacuated.

9:46 a.m.: Command Center notifies FAA headquarters that United 93 was 29 minutes away from Washington, D.C.

9:49 a.m.: FAA Command Center addresses Cleveland's 9:36 request to seek military intervention of United 93.

Command Center: *"Uh, do we want to think about, uh, scrambling aircraft?"*

FAA Headquarters: "Uh, God, I don't know."

Command Center: *"Uh, that's a decision somebody's gonna have to make probably in the next 10 minutes."*

9:57 a.m.: Bush departs Florida for Barksdale Air Force Base in Louisiana.

10:01 a.m.: Command Center tells FAA headquarters that another aircraft had seen United 93 "waving his wings." It's believed to be evidence of the passengers' efforts to overpower the hijackers.

10:03 a.m.: United Airlines Flight 93, en route from Newark, N.J., to San Francisco with 38 passengers, two pilots and five flight attendants aboard, crashes about 60 miles southeast of Pittsburgh.

10:05 a.m.: The south tower of the World Trade Center collapses.

10:10 a.m.: Part of the Pentagon collapses.

10:13 a.m.: The 39-story United Nations building is evacuated. A total of 11,700 people were evacuated "as a precautionary measure," a U.N. spokeswoman said.

10:24 a.m.: The FAA announces that all inbound trans-Atlantic aircraft into the United States are being diverted to Canada.

10:28 a.m.: The World Trade Center's north tower collapses from the top down.

10:45 a.m.: All federal office buildings in Washington are evacuated.

10:52 a.m.: Washington-area airports have been closed, a Federal Aviation Administration spokeswoman says.

10:53 a.m.: New York's primary elections scheduled for today are postponed.

11:02 a.m.: New York City Mayor Rudy Giuliani orders the evacuation of the area south of the World Trade Center.

11:18 a.m.: American Airlines says it has lost two planes. The airline says American Flight 11, a Boeing 767, en route from Boston to Los Angeles, and Flight 77, a Boeing 757 flying from Dulles Airport to San Francisco, had crashed.

11:26 a.m.: United Airlines announces the crash of United Flight 93 southeast of Pittsburgh.

11:59 a.m.: United Airlines confirms that Flight 175 has crashed with 56 passengers and nine crew members aboard.

1:04 p.m.: Bush, speaking from Barksdale Air Force Base, La., says the U.S. military has been put on high alert worldwide. He asks for prayers for those killed or wounded in the attacks. "Make no mistake, the United States will hunt down and punish those responsible for these cowardly acts," he said.

1:48 p.m.: Bush leaves Barksdale aboard Air Force One and flies to Offutt Air Force Base in Omaha, Nebraska.

5:20 p.m.: World Trade Center Tower 7 collapses.*

6:41 p.m.: Defense Secretary Donald Rumsfeld holds a news conference at the Pentagon. Rumsfeld announces the Pentagon will reopen for business today.

6:54 p.m.: Bush returns to the White House from Omaha.

7:15 p.m.: In a White House briefing Attorney General John Ashcroft said, "We will not tolerate such acts." He said the government had set up a Web site for tips, ifccfbi.gov.

7:25 p.m.: The congressional leadership holds a briefing on the steps of the U.S. Capital. Senate plurality leader Tom Daschle, D-South Dakota says Congress will reconvene this morning.

8:30 p.m.: President Bush addresses the nation from the Oval Office.

In the aftermath of the attack, it took four days before airline flights began again. Some crewmembers refused to fly because they were not confident adequate security was in place to protect them. With four 9/11 airliners hijacked and destroyed in a single day — you could hardly blame them.

Those who braved the return to work entered a very different workplace. No guidance from the airlines or the Federal Aviation Administration (FAA) awaited them about new or improved procedures to handle future such hijackings. Flight attendants inventoried galleys for anything they could use as a defensive weapon. Shell-shocked passengers often hugged flight attendants as they boarded. Some crewmembers barely contained tears, some hid in galleys to avoid alarming passengers. All the airline crews felt the attack on a more personal level than most. We had lost friends and co-workers that shared our world — it was devastating!

As airline crews returned to the skies, a new danger was posed — layoffs. On September 15, Continental Airlines released a statement that it would cut 12,000 jobs. One by one, other airlines followed suit: United and American announced 20,000 layoffs each; US Airways 11,000; Northwest, 10,000; Delta forecast pink slips for 13,000. The layoff toll topped 140,000. Washington Post writer David Montgomery quipped: "What thanks are flight attendants getting? How does this sound: You're fired."

As previously stated the 9/11 tragedy hit the airline industry hard, and it also gave an opening to accelerate airline restructuring already underway.

I was back flying a few days after 9/11. It was strange to see the terminals so empty and the planes even emptier. At the end of a three-day trip my last leg was from LAX to DTW on a Boeing 757. We had two people in first class and three in coach on a plane that could hold several hundred passengers. Our inflight rules about opening the cockpit door were still evolving, but let me suffice to say they were crude at best. As I remember the F/O would place the cockpit crash axe in his lap (yes we carry a razor sharp axe in the cockpit) and before a flight attendant could enter the flight deck another F/A would barricade the area in front of the door with galley cart — presumably to slow down any attempt to breach the cockpit. Then and only then could the F/A on the cockpit side of the barricade enter the flight deck. Needless to say getting a cup of coffee was an ordeal. Our bladders were equally abused.

After leveling off at cruise altitude it was over an hour before the lead F/A called the cockpit to see if we needed anything. Both the F/O and I were content, and honestly didn't mind keeping the door locked. I inquired who the first class passengers were and she said something like, "They are definitely not terrorist, we have comedians Tim Conway and Don Knotts sitting in First."

I grew up laughing at the on-screen antics of both of these gentlemen — they were two of the funniest people on the planet. The fact they were on board seemed strange, considering the fear most travelers had at the time. I asked the F/A if she'd invite the comics to the cockpit when we landed in Detroit — surprisingly they jumped at the chance.

It was nearing midnight when we landed in Detroit. Once at the gate the airplane emptied within seconds. After responding to the Parking Check List I opened the cockpit door and looked directly at Tim Conway and Don Knotts waiting patiently with big smiles, and outwardly excited to visit the cockpit. After a brief handshake I invited them to both climb up front and sit in the pilots seats. For me the rest is show biz legend. I spent the next 20 minutes entertained by the two funniest and 'real-people' celebrities I have ever encountered — and there have been many.

My second most memorable moment is captured in the photo I have with the two legendary (not afraid to fly) comics. The short time I spent with them was amazing. They are truly as funny in real life as they were on TV and the silver screen.

CHAPTER 17
The Right Stuff —
A Necessary Ingredient?

In 1979 author Tom Wolfe released a riveting book about pilots engaged in postwar experiments with experimental and rocket powered aircraft, These activities lead to the creation of NASA and ushered in the space age with 7 Mercury Astronauts making the first space flights for the U.S. (Russia was first into space with Cosmonaut Yuri Gagarin on April 12, 1961.) Shortly after release of "The Right Stuff," A movie deal was made and Wolfe's book moved to the silver screen. This is probably one of the best aviation films ever made — right up there with "Top Gun."

So what is the "Right Stuff?" Do all pilots have it, or do they even need it? The answer is probably yes... and no.

Being a test pilot, and flying an untested aircraft for the first time is akin to a non-swimmer being thrown into the deep end of the pool and told, "Either learn how to swim or you're going to die." Well, it's probably not that bad, but a large number of experimental test pilots have lost their lives on first flights and "Pushing the envelope," to coin a phrase from "The Right Stuff."

During a 'between airlines' period in my early flying life I was a research pilot on a classified government project. Not exactly a test pilot, but I routinely flew an experimental jet aircraft outside its normal flight envelope. Did I have or need the "Right Stuff?" Truthfully, I don't know. Pilots fly because they love to. We accept perceived or real risk in order to reap rewards, which can be as simple as watching the sunrise from 45,000 feet. Others love the rush of speed and sense of accomplishment in becoming a captain. The reasons are many.

Earning an Airline Transport Rating that's needed to captain a large (over 12,500 pound gross weight) aircraft is not simply a matter of passing a check ride. An FAA knowledge test, the most difficult and comprehensive aviation knowledge test around, must be passed before taking the check ride.

If you think you've got the "Right Stuff," try these questions on for size. If you want to know the correct answers, look them up — that's what the FAA wants you to do. It's their way of encouraging self-study.

The following is a reprint of some sample ATP test questions available at **www.faa.gov**. If you can answer 70 percent of these question correctly — you might have the right stuff.

---

1 . PLT103
When a recently certificated pilot decides to not wait any longer for the fog and low ceilings to lift, this pilot may be exhibiting the hazardous

  A) resigned attitude.
  B) macho attitude.
  C) impulsive attitude.

2 . PLT104
An air carrier aircraft flown into the ground while the crew is troubleshooting a landing gear fault is an example of

  A) neglect and reliance on memory.
  B) loss of situational awareness.
  C) lack of aviation experience.

3 . PLT103
Accident prone pilots tend to

  A) have disdain toward rules.
  B) follow methodical information gathering techniques.
  C) excessively utilize outside resources.

4 . PLT104

Automation has been found to

A) create higher workloads in terminal areas.

B) improve crew situational awareness skills.

C) substitute for a lack of aviation experience.

5 . PLT506

The maximum speed during takeoff that the
pilot may abort the takeoff and stop the airplane
within the accelerate-stop distance is

A) VEF.

B) V1.

C) V2.

6 . PLT395

What is the name of an area beyond the end of
a runway which does not contain obstructions
and can be considered when calculating takeoff
performance of turbine-powered aircraft?

A) Stopway.

B) Obstruction clearance plane.

C) Clearway.

7 . PLT432

Operational control of a flight refers to

A) exercising the privileges of pilot in command of an aircraft.

B) the specific duties of any required crewmember.

C) exercising authority over initiating, conducting, or terminating a flight.

8 . PLT493

What action is required prior to takeoff if snow is adhering to the wings of an air carrier airplane?

A) Add 15 knots to the normal VR speed as the snow will blow off.

B) Sweep off as much snow as possible and the residue must be polished smooth.

C) Assure that the snow is removed from the airplane.

9 . PLT409

You are traveling deadhead to or from a duty assignment. How does this affect the computation of flight time limits for air carrier flight crewmembers? It is

A) not considered to be part of a rest period.

B) considered part of the rest period for flight crew members.

C) considered part of the rest period if the flight crew includes more than two pilots.

10 . PLT444

Assuring that appropriate aeronautical charts are aboard an aircraft is the responsibility of the

A) first officer.
B) pilot in command.
C) aircraft dispatcher.

11 . PLT029

Except when in cruise flight, below what altitude are non-safety related cockpit activities by flight crewmembers prohibited?

A) FL 180.
B) 14,500 feet.
C) 10,000 feet.

12 . PLT436

If a required instrument on a multiengine airplane becomes inoperative, which document required under 14 CFR part 121 dictates whether the flight may continue en route?

A) A Master Minimum Equipment List for the airplane.
B) Certificate holder's manual.
C) Original dispatch release.

13 . PLT404

For a flight over uninhabited terrain, an airplane operated by a flag or supplemental air carrier must carry enough appropriately equipped survival kits for

A) all passenger seats.

B) all aircraft occupants.

C) all of the passengers, plus 10 percent.

14 . PLT449

A pilot in command operating under 14 CFR part 121 must complete a proficiency check or simulator training within the preceding

A) 24 calendar months.

B) 6 calendar months.

C) 12 calendar months.

15 . PLT436

Which 14 CFR part 121 required document includes descriptions of the required crewmember functions to be performed in the event of an emergency?

A) Airplane Flight Manual.

B) Pilot's Emergency Procedures Handbook.

C) Certificate holder's manual.

16 . PLT463

How soon after the conviction for driving while intoxicated by alcohol or drugs shall it be reported to the FAA, Civil Aviation Security Division?

A) No later than 60 days after the motor vehicle action.
B) No later than 30 working days after the motor vehicle action.
C) Required to be reported upon renewal of medical certificate.

17 . PLT463

A person may not act as a crewmember of a civil aircraft if alcoholic beverages have been consumed by that person within the preceding

A) 12 hours.
B) 24 hours.
C) 8 hours.

18 . PLT420

What minimum ground visibility may be used instead of a prescribed visibility criteria of RVR 16 when that RVR value is not reported?

A) 1/4 SM.
B) 1/2 SM.
C) 3/4 SM.

19 . PLT405

An approved minimum equipment list or FAA Letter of Authorization allows certain instruments or equipment to be inoperative

A) prior to beginning a flight in an aircraft if prescribed procedures are followed.

B) anytime with no other documentation required or procedures to be followed.

C) for a one time ferry flight of a large airplane to a maintenance base without further documentation from the operator or FAA with passengers on board.

20 . PLT147

A pilot approaching to land at a class D airport in a turbine-powered airplane on a runway served by a VASI shall

A) maintain an altitude at or above the glide slope until a lower altitude is necessary for a safe landing.

B) use the VASI only when weather conditions are below basic VFR.

C) not use the VASI unless a clearance for a VASI approach is received.

21 . PLT450

An example of air carrier experience a pilot may use towards the 1,000 hours required to serve as PIC in part 121 is flight time as an SIC

A) in part 121 operations.

B) in part 91, subpart K operations.

C) in part135 operations.

22 . PLT450

The holder of an ATP certificate with restricted privileges or an ATP certificate who also holds an aircraft type rating for the aircraft to be flown may act as

A) a PIC for a part 121 supplemental air carrier.

B) a PIC for a part 121 air carrier with 500 hours as a part121 SIC.

C) an SIC for a part 121 air carrier.

23 . PLT161

The maximum indicated airspeed that an aircraft may be flown in Class B airspace, after departing the primary airport, while at 1,700 feet AGL and 3.5 nautical miles from the airport is

A) 250 knots.

B) 200 knots.

C) 230 knots.

24 . PLT515

The Federal Aviation Administration's Flight Information Service Data Link (FISDL) provides what products?

A) METARs, SIGMETs, PIREPs, and AIRMETs.
B) Convective SIGMETs, PIREPs, AWWs, and NOTAMs.
C) SPECIs, SIGMETs, NOTAMs, and AIRMETs.

25 . PLT124

How does Vs (KTAS) speed vary with altitude?

A) Remains the same at all altitudes.
B) Varies directly with altitude.
C) Varies inversely with altitude.

26 . PLT523

Which is a purpose of wing-mounted vortex generators?

A) Delays the onset of drag divergence at high speeds and aids in maintaining aileron effectiveness at high angles of attack.
B) Breaks the airflow over the wings so the stall will progress from the root out to the tip of the wing.
C) Increase the onset of drag divergence and aid in aileron effectiveness at low speed.

27 . PLT473

Which is a purpose of ground spoilers?

A) Aid in rolling an airplane into a turn.

B) Increase the rate of descent without gaining airspeed.

C) Reduce the wings' lift upon landing.

28 . PLT170

Approaching the runway 1° below glidepath can add how many feet to the landing distance?

A) 250 feet.

B) 500 feet.

C) 1,000 feet.

29 . PLT134

One typical takeoff error is

A) delayed rotation which may extend the climb distance.

B) premature rotation which may increase takeoff distance.

C) extended rotation which may degrade acceleration.

30 . PLT303

What is the effect on total drag of an aircraft if the airspeed decreases in level flight below that speed for maximum L/D?

A) Drag increases because of increased parasite drag.

B) Drag decreases because of lower induced drag.

C) Drag increases because of increased induced drag.

31 . PLT347

Which engine is the 'critical' engine of a twin-engine airplane?

A) engine whose failure has the most adverse effect on directional control.

B) The engine whose failure has the least adverse effect on directional control.

C) The engine that is operating when used by the manufacturer to determine Vmc.

32 . PLT266

Swept wings causes a significant

A) increase in effectiveness of flaps.

B) reduction in effectiveness of flaps.

C) flap actuation reliability issue.

33 . PLT248

What result does a level turn have on the total lift required and load factor with a constant airspeed?

A) Lift required remains constant, and the load factor increases.

B) Both total lift required and load factor increase.

C) Lift required increases, and the load factor decreases.

34 . PLT248

What is the relationship of the rate of turn with the radius of turn with a constant angle of bank but increasing airspeed?

A) Rate will decrease and radius will increase.

B) Rate and radius will increase.

C) Rate will increase and radius will decrease.

35 . PLT237

By changing the angle of attack of a wing, the pilot can control the airplane's

A) lift, gross weight, and drag.

B) lift and airspeed, but not drag.

C) lift, airspeed, and drag.

36 . PLT477

The stall speed of an airplane

A) is constant regardless of weight or airfoil configuration.

B) is affected by weight and bank angle.

C) is not affected by dynamic pressures and lift coefficient.

37 . PLT519

What is a purpose of flight spoilers?

    A) Increase the camber of the wing.
    B) Direct air flow over the top of the
       wing at high angles of attack.
    C) Increase the rate of descent without
       increasing airspeed.

38 . PLT214

What is the result of a shock-induced separation of airflow occurring symmetrically near the wing root of a swept wing aircraft?

    A) A high-speed stall and sudden pitch up.
    B) Severe porpoising.
    C) A severe moment or 'Mach tuck.'

39 . PLT245

How can turbulent air cause an increase in stalling speed of an airfoil?

    A) A decrease in angle of attack.
    B) An abrupt change in relative wind.
    C) Sudden decrease in load factor.

40 . PLT094

The increase in specific range with altitude of the turbojet airplane can be attributed to three factors. One of those factors is

- A) an increase in altitude in the troposphere results in higher energy air flow.
- B) an increase in proportion of velocity versus thrust required.
- C) decreased engine turbine speeds.

41 . PLT213

Identify the type stability if the aircraft attitude tends to move farther from its original position after the controls have been neutralized.

- A) Negative static stability.
- B) Negative dynamic stability.
- C) Positive static stability.

42 . PLT140

A Land and Hold Short Operations (LAHSO) clearance, that the pilot accepts:

- A) does not preclude a rejected landing.
- B) precludes a rejected landing.
- C) must result in a landing.

43 . PLT171

What action should a pilot take if asked by ARTCC to "VERIFY 9,000" and the flight is actually maintaining 8,000?

    A) Immediately climb to 9,000.
    B) Report maintaining 8,000.
    C) Report climbing to 9,000.

44 . PLT083

Note: some of the following questions refer to charts and graphs that are part of the normal test package. If you're interested you can find the charts at: **http://www.faa.gov/training_testing/testing/test_questions/media/Addendum_C_ATP_Sup_7C.pdf**

For this book I've excluded those materials but included the questions as examples for you to contemplate.

(Refer to appendix 2, figures 255A, 255B, 256, 257 and 257A.) If the glide slope indication is lost upon passing LIMMA INT on the ILS RWY 25L approach at LAX, what action should the pilot take?

    A) Continue to the MAP, and execute the missed approach as indicated.
    B) Continue the approach as an LOC, and add 100 feet to the DH.
    C) Immediately start the missed approach left turn to CATLY INT.

45 . PLT149

As you rolled out long on Runway 30 after landing at Long Beach (LGB) (figures 241 and 242), you slowed and turned left on very wide pavement and now see Taxiway D signs on both sides of your pavement. You notice your heading is about 250°. Tower is urging you to turn left on D, cross 16R/34L, then taxi to G and hold short of Runway 30. You now know you

A) exited onto Runway 25R and transited HS 2.
B) exited ontoTaxiway G.
C) exited at Taxiway J and transited HS4.

46 . PLT370

What minimum information does an abbreviated departure clearance 'cleared as filed' include?

A) Clearance limit, transponder code,
and DP, if appropriate.
B) Destination airport, enroute altitude,
transponder code, and DP, if appropriate.
C) Clearance limit and enroute altitude.

47 . PLT195

Each pilot who deviates from an ATC clearance in response to a TCAS II, resolution advisory (RA) is expected to

A) maintain the course and altitude resulting from
the deviation, as ATC has radar contact.
B) notify ATC of the deviation as soon as practicable.
C) request ATC clearance for the deviation.

48 . PLT058

(Refer to appendix 2, figure 171, top panel.) The facility (Kankakee) that is located 9 miles NE of Chicago Midway or 27 miles SSE of Northbrook (OBK) is a/an

A) Aeronautical Radio Inc. (AIRINC) transmitter.
B) Flight Service, Remote Communications Outlet.
C) Automated Weather Observing System (AWOS/ASOS) with frequency.

49 . PLT149

What special consideration is given for turbine-powered aircraft when 'gate hold' procedures are in effect?

A) They are expected to be ready for takeoff when they reach the runway or warm up block.
B) They are expected to be ready for takeoff prior to taxi and will receive takeoff clearance prior to taxi.
C) They are given preference for departure over other aircraft.

50 . PLT362

You notice ATC is unusually quiet and one of your VHF transmit lights is illuminated, you suspect

A) your VHF receiver is inoperative.
B) your VHF transmitter is keyed and you probably have a stuck microphone.
C) the radio is performing a self-test function.

51 . PLT141
Taxiway Centerline Lead-Off Lights are
color coded to warn pilots that

A) they are within the runway environment
   or run-up danger critical area.
B) they are within the runway environment or ILS
   critical area.
C) they are within the taxiway end
   environment or ILS critical area.

52 . PLT141
Runway Status Lights (REL) are

A) an independent light system.
B) automatically activated.
C) ATC tower controlled.

53 . PLT149
You received these ATC taxi instructions: "Taxi to Runway
30 via Lima and hold short of Runway 25L." Your airplane
is on the ramp by the terminal and NWS on the east
side of the airport. (See figure 242.) Your taxi route

A) requires crossing of Runway 25L at Lima.
B) involves transiting HS4.
C) requires crossing Runway 34R enroute to the
   assigned runway.

54 . PLT149
When taxiing on an airport with ASDE-X, you should

A) operate the transponder only when the airport is under IFR or at night during your taxi.

B) operate the transponder with altitude reporting all of the time during taxiing.

C) be ready to activate the transponder upon ATC request while taxiing.

55 . PLT367
Before requesting RVSM clearance, each person

A) shall correctly annotate the flight plan.

B) must file an ICAO RVSM flight plan.

C) should file for odd altitudes only.

56 . PLT061
KFTW UA/OV DFW/TM 1645/FL100/TP PA30/
SK SCT031-TOP043/BKN060-TOP085/OVC097-
TOPUNKN/WX FV00SM RA/TA 07.

This pilot report to Fort Worth (KFTW) indicates

A) the aircraft is in light rain.

B) the ceiling at KDFW is 6,000 feet.

C) that the top of the ceiling is 4,300 feet.

57 . PLT076
(Refer to appendix 2, figure 149.) What will be the wind and temperature trend for an SAT ELP TUS flight at 16,000 feet?

A) Temperature decrease slightly.

B) Wind direction shift from southwest to east.

C) Wind speed decrease.

58 . PLT147

A pilot of a high-performance airplane should
be aware that flying a steeper-than-normal
VASI glide slope angle may result in

A) a hard landing.
B) landing short of the runway threshold.
C) increased landing rollout.

59 . PLT144

What effect, if any, will landing at a higher-than-
recommended touchdown speed have on hydroplaning?

A) Increases hydroplaning potential regardless of braking.
B) No effect on hydroplaning, but increases landing roll.
C) Reduces hydroplaning potential if heavy braking
   is applied.

60 . PLT008

(Refer to FAA-CT-8080-7C, Addendum B, Figure 331
and Addendum C, Figure 461.) At a weight of 73,500
pounds, the expected Landing Field Length is

A) 6,700 feet.
B) 5,700 feet.
C) 6,500 feet.

61 . PLT123
(Refer to FAA-CT-8080-7C, Addendum C, Figure 465.) At a weight of 60,000 pounds with 35° flaps, the reference stall speed is

A) 96 knots.
B) 124 knots.
C) 101 knots.

62 . PLT089
(Refer to FAA-CT-8080-7C, Addendum B, Figure 321 and Addendum C, Figure 471.) With a reported temperature of -5°C and gross weight of 49,000 pounds, the chart V2 value is

A) 118 knots.
B) 120 knots.
C) 122 knots.

63 . PLT011
(Refer to FAA-CT-8080-7C, Addendum B, Figure 297 and Addendum C, Figure 478.) With a reported temperature of 25°C, a weight of 55,000 pounds, and a V1/VR ratio of 0.95, the accelerate-stop distance required is

A) 5,500 feet.
B) 4,300 feet.
C) 5,900 feet.

64 . PLT121
(Refer to FAA-CT-8080-7C, Addendum B, Figure
321 and Addendum C Figure 458.) With a reported
temperature of 15°C, a 0.8% upslope, and
calm winds, the maximum permissible
quick turn-around landing weight is

   A) 80,700 pounds.
   B) 72,500 pounds.
   C) 84,000 pounds.

65 . PLT011
(Refer to FAA-CT-8080-7C, Addendum B, Figure
363 and Addendum C, Figure 429.) At a reported
temperature of 10°C with cowl anti-ice on and
packs on, the takeoff thrust setting is

   A) 90.0%.
   B) 89.1%.
   C) 87.4%.

66 . PLT013
(Refer to FAA-CT-8080-7C, Addendum B, Figure 287 and
Addendum C, Figure 421.) The winds are reported as
220/15. You compute the tailwind component hoping for
a Runway 33 takeoff. You compute the tailwind to be

   A) 14 knots.
   B) 10 knots.
   C) 5 knots.

67 . PLT089
(Refer to FAA-CT-8080-7C, Addendum B, Figure 340 and
Addendum C, Figure 450.) With a reported temperature
of 35°C, flaps set at 8, and 5 knots of headwind at a
takeoff weight of 82,300 pounds, the V1MBE is

    A) 174 knots.
    B) 171 knots.
    C) 142 knots.

68 . PLT004
(Refer to FAA-CT-8080-7C, Addendum B, Figure 273 and
Addendum C, Figure 474.) With a reported temperature
of 45°C and a weight of 52,000 pounds, the first segment
one engine inoperative takeoff gross climb gradient is

    A) 0.020%.
    B) 0.043%.
    C) 0.032%.

69 . PLT123
(Refer to FAA-CT-8080-7C, Addendum C, Figure 466.) At a
weight of 60,500 pounds with 5° flaps, the 1.3 VSR speed

    A) 159 knots.
    B) 148 knots.
    C) 163 knots.

70 . PLT004
(Refer to FAA-CT-8080-7C, Addendum C,
Figure 472.) With a gross weight of 54,500
pounds, the final takeoff climb speed is

A) 142 knots.
B) 145 knots.
C) 148 knots.

71 . PLT011
(Refer to FAA-CT-8080-7C, Addendum B, Figure
297 and Addendum C, Figure 481.) With a reported
temperature of 0°C, at 500 feet AGL after takeoff, and
an airspeed of 145 knots IAS, the radius of turn is

A) 6,650 feet.
B) 8,000 feet.
C) 9,700 feet.

72 . PLT004
(Refer to FAA-CT-8080-7C, Addendum B, Figure
273 and Addendum C, Figure 475.) With a reported
temperature of 32°C, and a weight of 58,000 pounds,
the second segment takeoff gross climb gradient is

A) 0.057%.
B) 0.062%.
C) 0.034%.

73 . PLT008
(Refer to FAA-CT-8080-7C, Addendum C, Figure 460.) At a weight of 77,500 pounds, and a landing elevation below 5,000 feet, the VRef is

A) 139 knots.
B) 141knots.
C) 143knots.

74 . PLT121
(Refer to FAA-CT-8080-7C, Addendum C, Figure 459.) For a supplemental charter, a still air range of 2,250 NM is required. The payload for this non-stop trip is

A) 5,100 pounds.
B) 5,600 pounds.
C) 6,100 pounds.

75 . PLT499
Which part(s) in the turbojet engine is subjected to the high temperatures and severe centrifugal forces?

A) Turbine wheel(s).
B) Turbine vanes.
C) Compress or rotor(s) or impeller(s).

76 . PLT500
Equivalent shaft horsepower (ESHP) of a turboprop engine is a measure of

A) turbine inlet temperature.
B) propeller thrust only.
C) shaft horsepower and jet thrust.

77 . PLT499

The most important restriction to the operation
of turbojet or turboprop engines is

    A) limiting compressor speed.
    B) limiting torque.
    C) limiting exhaust gas temperature.

78 . PLT127

As outside air pressure decreases, thrust output will

    A) remain the same since compression of inlet air will
        compensate for any decrease in air pressure.
    B) increase due to greater efficiency
        of jet aircraft in thin air.
    C) decrease due to higher density altitude.

79 . PLT104

When a pilot who is new to advanced avionics operations
operates closer to personal or environmental limits,

    A) greater utilization of the aircraft is achieved.
    B) risk is increased.
    C) risk is decreased.

80 . PLT280

Sudden penetration of fog can create the illusion of

    A) leveling off.
    B) pitching up.
    C) pitching down.

81 . PLT104

The lighter workloads associated with glass (digital) flight instrumentation

    A) are instrumental in decreasing training requirements.
    B) have proven to increase basic flight skills.
    C) may lead to complacency by the flight crew.

82 . PLT097

What is a symptom of carbon monoxide poisoning?

    A) Rapid, shallow breathing.
    B) Dizziness.
    C) Pain and cramping of the hands and feet.

83 . PLT205

What is the effect of alcohol consumption on functions of the body?

    A) Alcohol has an adverse effect, especially as altitude increases.
    B) Alcohol has little effect if followed by an ounce of black coffee for every ounce of alcohol.
    C) Small amounts of alcohol in the human system increase judgment and decision-making abilities.

84 . PLT332

Which is a common symptom of hyperventilation?

    A) Increased vision keenness.
    B) Decreased breathing rate.
    C) Tingling of the hands, legs, and feet.

85 . PLT280

The illusion of being in a nose up attitude which may occur during a rapid acceleration takeoff is known as

A) somatogravic illusion.
B) autokinesis.
C) inversion illusion.

86 . PLT104 Human behavior

A) rarely results in accidents unless
   deliberate actions are performed.
B) is responsible for three out of four accidents.
C) is well understood, so behavioral induced
   accidents are exceedingly rare occurrences.

87 . PLT512 Large areas of land

A) tend to increase temperature variations.
B) do not influence the troposphere.
C) minimize temperature variations.

88 . PLT511

What is a feature of a stationary front?

A) Weather conditions are a combination of strong
   cold front and strong warm front weather.
B) The warm front surface moves about half
   the speed of the cold front surface.
C) Surface winds tend to flow parallel to the frontal zone.

89 . PLT203
Which feature is associated with the tropopause?

A) Absence of wind and turbulence.
B) Abrupt change of temperature lapse rate.
C) Absolute upper limit of cloud formation.

90 . PLT475
Where do squall lines most often develop?

A) Ahead of a cold front.
B) In an occluded front.
C) Behind a stationary front.

91 . PLT495
Convective clouds which penetrate a stratus layer
can produce which threat to instrument flight?

A) Freezing rain.
B) Embedded thunderstorms.
C) Clear air turbulence.

92 . PLT475
If squalls are reported at the destination
airport, what wind conditions exist?

A) Sudden increases in wind speed of at least
   15 knots to a sustained wind speed of 20
   knots, lasting for at least 1 minute.
B) Rapid variation in wind direction of at least 20° and
   changes in speed of at least 10knots between peaks.
C) A sudden increase in wind speed of at least 16 knots,
   the speed rising to 22 knots or more for 1 minute
   or more.

93 . PLT302

Which type clouds may be associated with the jetstream?

A) Cumulonimbus cloud line where the
   jetstream crosses the cold front.
B) Cirrostratus cloud band on the polar
   side and under the jetstream.
C) Cirrus clouds on the equatorial side of the jetstream.

94 . PLT302

Where are jetstreams normally located?

A) In a break in the tropopause where intensified
   temperature gradients are located.
B) In areas of strong lowpressure
   systems in the stratosphere.
C) In a single continuous band, encircling the
   Earth, where there is a break between
   the equatorial and polar tropopause.

95 . PLT493

Which conditions result in the formation of frost?

A) The temperature of the collecting
   surface is at or below freezing and small
   droplets of moisture are falling.
B) Temperature of the collecting surface is below the
   dewpoint and the dewpoint is also below freezing.
C) Dew collects on the surface and then freezes because
   the surface temperature is lower than the air.

96 . PLT301

What characterizes a ground-based inversion?

A) Cold temperatures.
B) Poor visibility.
C) Convection currents at the surface.

97 . PLT108

Freezing Point Depressant (FPD) fluids used for deicing

A) on the ground, cause no performance
degradation during takeoff.
B) provide ice protection during flight.
C) are intended to provide ice protection on the
ground only.

98 . PLT274

When you hear a SIGMET on an ATC frequency
forecasting severe icing conditions on the
route to your destination, you plan for

A) the installed transport category airplane
ice protection system protecting against all
types and levels of icing as designed.
B) very little airframe icing because of an OAT of-
10°Corcolder, the moisture is already frozen
and cannot adhere to airplane surfaces.
C) the possibility of freezing rain and freezing drizzle
that can accumulate on and beyond the limits of
any aircraft.

99 . PLT047

When using a flight director system, what rate of turn or bank angle should a pilot observe during turns in a holding pattern?

    A) 3° per second or 25° bank, whichever is less.
    B) 1-1/2°per second or 25°bank,whichever is less.
    C) 3° per second or 30° bank, whichever is less.

100 . PLT128

During an en route descent, both the ram air input and drain hole of the pitot system becomes completely blocked by ice. What airspeed indication can be expected?

    A) Increase in indicated airspeed.

    B) Indicated airspeed remains at the value prior to icing.

    C) Decrease in indicated airspeed.

CHAPTER 18
Probation — Making The Cut

This book ends on a note of finality and hopefully recaps the incredible journey to the captain's seat for those seeking a very alternative lifestyle.

For the casual reader, I hope you have a new found respect for those men and women who wear the stripes of an air crew, especially the legions of flight attendants who make a pilot's job easier and provide an incredible safety net for our passengers: you all have my undying respect and admiration.

For those few individuals who actually pursue and are lucky enough to land a pilot job with a major airline consider yourself under the looking glass for at least a year. These are my parting words and warnings to you.

In addition to the many other responsibilities captains have they also are the most reliable resource for finding out about the adaptability of new hire pilots. Throughout the first year of a pilot's employment they will likely fly with dozens of different captains. Each of these captains may be asked for a written review of the new hire's performance on the job. These reviews typically go to the chief pilot's office at the base where the pilot is assigned. Periodically the probation pilot will be called to the chief pilot's office for review of their performance.

Over the years I have filled out dozens of these evaluation forms. In my opinion it is one of the most import duties I had as a captain. Here's why.

I've found that most people have the ability to show their best side when it's required. Often times this 'best side' is not who they really are, but what they think others wish them to be. Nowhere is this more evident than when you're flying with a new hire. You're on a two to five day trip with someone you've just met. That person has the ability to take your dream job away from you if you fail to impress him or her. Naturally you're going to try and show the your best work as a pilot and team player. But is it an act, or is it really who you are? The captain's job is to evaluate you based on what they observe, not necessarily on what they think.

The evaluation report is typically a form filled with relevant questions on timeliness, attention to detail, preparedness and piloting skills. For the sake of expediency for the captain these are usually answered by check marks in satisfactory or unsatisfactory boxes. The probationary pilot is often times the one who hands the form to the captain. Typically new hires are required to turn in at least one evaluation form each month. As you'd expect, some new-hires make sure all is going well between them and their captain before asking them to fill out an evaluation form. Some captains treat this lightly, but most are very serious about the process. The question I always asked myself about a new hire is whether this is an individual I can entrust my life with. It's not a simple question when you consider what's at stake.

If you're lucky the career as a F/O then captain will be one without incident. Unfortunately statistics show that pilots are in one of the most hazardous jobs on earth. In a recent ranking chart of the most dangerous professions, pilots and flight engineers ranked third –– NASCAR drivers are not even in the top 10.

I digress. When the manure hits the fan in a two pilot aircraft, doing your job correctly and quickly can mean the difference between life and death. Like most captains who've evaluated new hires I wanted to know if the other person could be counted on in times of crisis. This is mostly a 'gut' call since you seldom, if ever, get to see someone perform under an emergency situation other than in a simulator. Trust me, you can't simulate the fear that shrouds your thoughts when a real in-flight emergency threatens your life as well as those of hundreds of passengers.

A trip in smooth weather with no mechanicals is typical for about 99.9 percent of all airline flying, so it's difficult to gauge many of a pilots skills during such routine flights. By the time most new hires reach the airline cockpit a combination of Human Resource professionals, instructors and check airmen staff have already weighed in on a their suitability. Occasionally someone squeaks by and makes it to the line so now it's the line captain's job to make sure the first 12 months of a new pilot's career is monitored carefully. Like I said previously, the vetting process for a pilot job is a long one.

I had the fortune (or misfortune) of flying for five different airlines (three regional carriers and two majors) in my 35-year career. I changed jobs because I wanted to end up flying for a larger carrier. This is not necessarily a normal career path, but many are similar. The job changes during my first seven years of airline flying positioned me where I wanted to be, but also came at a hefty price. I had to endure five different year-long probations for five different airlines. Trust me, I'm an authority on the probationary year. As a consequence I loved flying with probationary pilots. I knew how they felt and could relate to how they acted in front of me.

Once your initial training is complete, you're qualified to fly with any captain assigned to the same equipment. The exception is newly minted captains are not normally paired with 'new-to-the-airplane' first officers until they have acquired some additional flight time in the new jet. This prevents two pilots that are new to an airplane (or seat position) from flying together until they've gained additional experience flying with more seasoned aviators on that equipment. It's a good system, but it wasn't always that way. Once upon a time you might have had a brand new cockpit crew up front with a minimum amount of experience in that equipment. However, it still happens to some degree when a brand new type of airplane is introduced to the airline. Instructor pilots who've gone through intense training as the initial pilot group typically captain those flights while training other crew members.

If you're at a major airline you can be initially trained on virtually any airplane they fly during your first year. You're on probation pay. This is usually an annual fixed salary, so regardless of the equipment you qualify on, your pay rate is the same as every other probation pilot.

In the past it was not uncommon for an airline to assign you to a senior piece of equipment (if the contract allows them to do so) this allows them to take advantage of the cheap labor on the normally high paying-larger jets. Over the years union contracts have evolved, and now most pilot contracts minimize this practice. Regardless, during a pilot's first year, the airline owns your uncomplaining, always punctual, professional self for 365 days. It's like indentured servitude.

Here's an added bonus to your first year. Most likely much of, if not all of, your probation period will likely be spent on reserve.

If you choose to commute to your base from out of town, be prepared to stay in a commuter pad, (hotel room shared by other crew or a rented house or apartment also shared by other pilots and sometimes flight attendants.) These quarters are typically called "Crash pads," which seems humorous considering pilot and flight attendant occupations.

Commuting to work is expensive, but this type of job only requires you show up for flights on time. Where you live is normally of no concern to the airline, although they love it when you live in base. The ability to commute is, by some measure, one of the pluses about airline flying, but it's also a huge minus in your checkbook. On reflection, my desire to live away from base for most of my career has probably cost me thousands and thousands of non-deductible-out-of-my-own-pocket dollars.

My advice to anyone looking for a career at the majors is to consider their crew bases before submitting an application. Living at your base is a huge cash and time saving measure. If you're inclined to be in pilot management or desire to become a check airman/ instructor-pilot, living at your base will also put you a lot closer to the action.

* * *

After 365 days of 'call-out' reserve flying in the middle of the night, cheap fast food on the run, trying to explain comments on your evaluation forms to the chief pilot, and sweating flights with a "Captain

Ahab," you're finally called in for an end of probation 'chat' with the base chief pilot or director of flight operations. This is usually a cursory review of your past years performance. On occasion it's also a "Goodbye, sorry it didn't work out," moment. The truth is, not everyone makes it, even after a year of miserable probation. These moments are usually handled delicately, but on occasion I have seen grown men leave the chief pilots office with tears in their eyes — never to be seen again.

Next time you see a pilot standing at the doorway saying goodbyes take a moment to shake their hand and say thank you for a safe trip. It's what we get paid to provide you and we worked hard for the privilege — besides, we still like knowing you care.

CHAPTER 19

The Last Landing —
Retirement, The Afterlife

As my retirement day approached I had mixed feelings about the 'after' life. Flying big jets around for a living, in my humble opinion, is probably the most gratifying long-term occupation on earth. Sure flying the shuttle or being a fighter pilot are exciting endeavors, but they are very short term.

My wife wanted to know if I wanted a party or her to accompany me on the last trip. I chose none of this, preferring to leave, just as I initially came into the airline industry — quietly and unnoticed.

Packing for the last trip I reflected on my youth — watching airplanes fly directly over my house on approach to San Diego's Lindbergh Field. Nothing else mattered to me but airplanes — the desire to pilot one of these gravity-defying machines was all consuming. At age 12 I requested applications and requirements for a pilot's position at United, Pan Am and TWA. As I remember they all replied with encouraging letters — mostly telling me to stay in school and try to fly in the military. I did all that and, lo and behold, I finally got the job.

Checking in for my last trip, I tried to downplay my quiet exit, but the check-in-clerk I had known for years actually came out to the plane with a tearful hug and goodbye as I sat slightly embarrassed in the captains seat. When she left the cockpit my F/O said, "What's up with that, boss?" Reluctantly I informed him of the situation — but asked that he keep the fact quiet. He said yes, but he meant no.

It was a three-day trip with an overnight in San Francisco on day two. Here's where the secrecy of my retirement disappeared.

After arrival at San Francisco the crew and I gathered up our bags and left the terminal through the security check point.

So far no one had mentioned my retirement.

As soon as we passed security I heard a female voice say, "Hey captain, want a date?" The voice echoed across the passenger/TSA chatter as I zeroed in on the source. I was surprised to see the wife of one of my close friends smiling as she looked directly at me. Debbie was a flight attendant for American and her husband Dick, a Boeing 767 captain for the same airline. I laughed returning her smile — unsure if this was a wife loan or chance meeting. As it turned out it was a little of both. The couple had somehow found out about my impending retirement and had decided to be there on my last night as a captain. She had flown in from Dallas to meet me — later that evening her husband arrived after captaining an international flight from Europe to Dallas. After his grueling workday he immediately hopped on a flight to SFO to join us. Now that's what you call a class act.

The evening was lovely with a wonderful dinner and gift of a luxurious blue bathrobe with four gold stripes sewn on each sleeve, and the name "Captain Jack Watson" on the breast. It sort of makes me look like a (since departed) Supreme Court Justice. Thanks Dick and Debbie for the wonderful memory.

The last leg home was from MSP to DTW. It was not my turn to fly the leg, but the F/O kindly said, "You got this one boss — make it count."

It was a cool crisp late afternoon as I rolled on the nearly full Boeing 757-300 for the last time. The "Landing-Gods," must have been with me. I never felt the wheels touch the ground. I rolled the entire length of the runway savoring these last moments. I was one with the plane — we were a single unit performing a ballet I had practiced all my adult life. As I turned off at the end of DTW runway 21 Right, my eyes moistened behind my Ray Ban's. That was the last time I would ever land a commercial airliner — the reality was sinking in.

The taxi from the end of the runway to the gate was the longest I ever remember. Flashbacks of thousands of flights coursed through my thoughts; some funny, some heartbreaking. Each of those countless trips taught me lessons about myself, but more importantly they brought me close to many talented people I was honored to fly and work with. As the arena of gates, planes and ground equipment swept behind me I eased the 250,000-pound airliner into my final gate.

The crew presented me with a signed goodbye card and as the 200 plus passengers exited the plane I shook hands with each and every one saying, "Thank you for the privilege," — and indeed it had been.

I now write books, have photographed models for Playboy enterprises, and my wife, four cats and I spend almost half of each year touring America in our motorhome.

While not necessarily as exciting, life after flying can be very rewarding — even if you never get the chance to photograph a Playboy model.

If after reading this book you decide a captain's life is the life for you. Remember these words. Be humble, gracious, courteous and kind, but most of all — patient. A time will come when you too can pick up the mic and say, "This is your captain!"

The End

AIRCREW SPEAK

Over many years of commercial air travel a universal and often cryptic language evolved among pilots and flight attendants. This language is used routinely and is often overheard or directed at passengers. For the curious traveler or aspiring pilot here's a short list of commonly used aircrew terms and their definitions.

ARM THE DOORS – Pre-flight task to assure door is secure and evacuation slides are armed.

A/C – Crew-speak for the aircraft.

ADD-ON – Normally refers to a flight attendant who was not listed on the flight schedule but was penciled in due to another calling in sick.

AIR POCKET -- Transient jolt of turbulence.

ALL-CALL – Often part of the arming/disarming procedure, a request that each flight attendant report via intercom from his or her station — flight attendant conference call.

ALLEY - A taxiway or passageway between terminals or RAMPS.

APRON - Similar to RAMP, any expanse of TARMAC not a runway or taxiway — i.e. areas where planes park or are serviced.

AREA OF WEATHER - This typically means thunderstorms or a zone of heavy precipitation.

BASE – Flight crew term for their home airport; where their flights originate and terminate.

BID – When a pilot or flight attendant puts in a request for a specific route or schedule, they are bidding. Usually done on a monthly basis.

BOTTLE TO THROTTLE – Curfew hours. In airline parlance, the time a person must abstain from having another alcoholic beverage and the time of their next flight; often 12 hours.

COMMUTE – Process of getting to starting destination or base. Pilots and flight attendants often live in one city but have another city as their base, and thus commute to work.

CRASH PAD – The term used by flight attendants and pilots in reference to a rented apartment, motel room or house usually shared by several flight attendants and or pilots.

CROTCH WATCH – Refers to rounds flight attendants make prior to liftoff and descent that ensures all passengers' seat belts are on and properly fastened.

CRUMB CRUNCHERS – Term flight attendants use in reference to children passengers.

DEADHEAD – When a pilot or flight attendant flies as a passenger, they are deadheading if it is part of their job.

DEPLANE - Deplane is used to describe the opposite of boarding an aircraft.

DIRECT FLIGHT – A direct flight is a routing along which the flight number does not change; it has nothing to do with whether the plane stops.

EFC TIME - Expect further clearance (EFC) time, sometimes called a release time, this is the point in time where a crew expects to be set free from a HOLDING PATTERN or a GROUND STOP.

EQUIPMENT - An airplane.

F/A – Flight attendant.

FERRY FLIGHT – Flight of an aircraft to a specific destination without paying passengers aboard.

FINAL APPROACH - An airplane is on final approach when it has reached the last, straight-in segment of the landing pattern that is, aligned with the extended centerline of the runway.

FIRST OFFICER (also, CO-PILOT) - Second in command on the FLIGHT DECK. They are fully qualified to operate the aircraft in all stages of flight, including takeoffs and landings.

FLIGHT DECK - The cockpit.

FLIGHT LEVEL - How many thousands of feet you are above sea level. Just add a couple of zeroes. Flight level three-three zero is 33,000 feet.

GEORGE – A colloquial term for "autopilot."

GREEN AIRCRAFT – Slang term referring to a plane fresh from the factory with exterior paint and interior still isn't complete.

GROIN SCAN – Another name for the "crotch watch," which is when the flight attendants make their rounds to ensure that all seat belts are properly fastened.

GROUND STOP - The point when departures to one or more destination are stopped by ATC, usually due to air traffic congestion at the destination airport.

HOLDING PATTERN - A racetrack-shaped course flown during weather or traffic delays.

ILLEGAL – Pilot or flight attendant, who crosses over the maximum hours allowed to work per flight, day, or schedule without a required sleep or break period.

IN RANGE - Somewhere around the start of descent, pilots send an electronic "in range" message to let everybody at the arrival station know they'll be arriving shortly.

INTERPHONE – The phones located throughout the cabin, allowing flight attendants to speak with each other and the flight deck when the cockpit door is shut.

JUMP SEAT – A flight term referring to an auxiliary (extra) seat for persons who are not operating the aircraft, such as the cabin crew or perhaps a trainee or deadheading pilot in the flight deck jump seat. Often used to denote the drop-down seat flight attendants strap themselves into during takeoff and landing.

KETTLE CLASS – A mocking term for economy or coach class, taken from the hillbilly characters of Ma and Pa Kettle.

LANDING LIPS – A term originally referring to when female flight attendants would put on their lipstick and other makeup to make themselves presentable to passengers as they bid them farewell.

LAST MINUTE PAPERWORK - Usually refers to last minute weight-and-balance corrections, a revision to the flight plan, or waiting for maintenance to deal with a write-up and get the logbook in order.

LINE – The schedule of trips a pilot or flight attendant is awarded after bid results are posted each month, referring to the entire sequence.

LINE HOLDER – A pilot or flight attendant senior enough to hold a line, and not on reserve.

LOUNGE LIZARD – A pilot or flight attendant who would skip the use of a crash pad; choosing instead to sleep in the crew lounge at the airport, usually to save money.

NARROWBODY – A plane with one aisle down the center. As opposed to a widebody.

NON-REV – Short for "Non-revenue," is an airline employee or a family member who takes advantage of the company perks to fly at a nominal cost, or free.

NONSTOP - Flight that doesn't stop.

OFFLINE CITY – A city the airline or flight does not normally fly to.

PAX – An abbreviation for "passengers."

PRE-BOARD – A term referring to those passengers who board first, and ahead of all first, business and coach class passengers.

RAMP - Ramp refers to the aircraft and ground vehicle movement areas closest to the terminal — the aircraft parking zones and surrounds.

SLAM-CLICK – In general slang, this means to go straight to one's room and lock the door (slam the door, click it shut/locked).

TARMAC – Refers to any ramp, apron, or taxiway surface.

TRANSCON – A transcontinental flight – across one continent or country.

TURN – A flight that leaves base and returns back to base in the same day. Also known as a turnaround.

WHEELS-UP TIME - Similar to the EFC TIME, except it refers to the time when a ground-stopped plane is expected to be fully airborne.

WIDEBODY – A large plane with a row of seats going down the center, meaning there are two aisles in the main cabin. As opposed to a narrowbody.

Z Time – Pilot-speak for Greenwich Mean Time.

ABOUT THE AUTHOR

Jack Watson is a critically acclaimed author, photographer, producer and filmmaker. He is a former military pilot and international airline Captain with over 32,000 hours' flight time, and almost a half-century of aviation experience. He is qualified in over 100 airplanes and a few helicopters. In addition to holding an Airline Transport Rating in both Fixed Wing and Rotor-wing aircraft he holds eight type ratings in jet transports. His other FAA ratings include Airframe and Powerplant mechanics license and a Flight Engineer Rating in both Turboprop and Turbojet transport category aircraft.

With over 35 years experience in the visual and literary arts, Watson's books, movies, photos and short stories have appeared worldwide.

Other Jack Watson books by Atlantic Publishing include:

DRONE WARRIOR

DRONE ACE

BEYOND THE GLAMOUR PHOTOGRAPH

GREEN SCREEN GLAMOUR

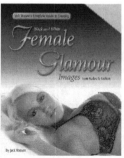

BLACK AND WHITE FEMALE GLAMOUR IMAGES FROM NUDES TO FASHION

HIDDEN CUBA: A PHOTOJOURNALIST'S UNAUTHORIZED JOURNEY TO CUBA TO CAPTURE DAILY LIFE: 50 YEARS AFTER CASTRO'S REVOLUTION

Jack's Hidden Cuba book received a prestigious GOLD MEDAL AWARD from the Florida Book Publishers Association and is a must have with breathtaking behind-the-scenes images of Cuba.

These books by Atlantic Publishing Group can be purchased through leading book retailers everywhere.

INDEX

G

H

I

N

P